Farm

Nicola Harvey was born in Aotearoa New Zealand, and called Australia home for almost two decades. She has written and edited for magazines and websites including *North & South*, *The Saturday Paper*, *frieze*, *The Listener*, BuzzFeed, and the ABC. Nicola has created and produced acclaimed podcast series including *Pretty For An Aboriginal* and *Debutante* in collaboration with Nakkiah Lui and Miranda Tapsell, and *A Carnivore's Crisis* with Rachel Khoo. She currently runs her own production company, Pipi Films, from her home office on a farm in Aotearoa. *Farm* is her first work of nonfiction.

Farm

the making of
a climate activist

Nicola Harvey

SCRIBE
Melbourne • London

Scribe Publications
18–20 Edward St, Brunswick, Victoria 3056, Australia
2 John St, Clerkenwell, London, WC1N 2ES, United Kingdom
3754 Pleasant Ave, Suite 100, Minneapolis, Minnesota 55409, USA

First published by Scribe in Australia 2022
Published by Scribe in North America 2023

Typeset in Adobe Caslon Pro by the publishers

Printed and bound in the UK by CPI Group (UK) Ltd, Croydon CR0 4YY

Scribe is committed to the sustainable use of natural resources and the use of paper products made responsibly from those resources.

978 1 957363 46 2 (US edition)
978 1 922310 54 5 (Australian edition)
978 1 925938 91 3 (ebook)

Catalogue records for this book are available from the National Library of Australia.

scribepublications.com
scribepublications.com.au
scribepublications.co.uk

For Pat, my great love

&

For Clara, who fuels my hope

Contents

Preface

I've been thinking about an artwork that I looked at for a long time when I saw it in an exhibition. It's a tiny tin-and-aluminium sculpture, created by Australian artist Fiona Hall, titled *Karra–wari (Pitjantjatjara) / Eucalyptus microtheca / Coolibah tree*, 1999. It's part of the series 'Paradisus Terrestris 1989–90'.

Depending on my mood, the artwork offers either a vision of hope or destruction. A sardine can has its lid rolled back to reveal a low relief sculpture of a male torso, side on, from armpit to groin. Sprouting from the top of the body is a delicate hand-cut tin coolibah tree, a eucalyptus found in western New South Wales where Pat, my husband, is from. The tin can has been cut, ripped, beaten, destroyed, and remade into a thing of startling beauty. The human form is either the potent life source for the tree or the thing buried to give way to nature.

In the eye of the climate change storm, we are all in that position. Are we to reach skyward for something that gives us life or fall beneath the surface, forever?

Introduction

I ate corn chips during my miscarriage. Sitting cross-legged on the shedding cream carpet in front of my double bed as dappled sunlight flowed and receded through the sheer linen IKEA curtains, I sucked the salt and tangy flavouring off each chip and crunched through the waves of pain. Hours later, I curled on the bed and closed my eyes. A rumble of hunger was the only thing cutting through my numbness. Next door, in the spare bedroom, my husband, Pat, lay on the floor, his head propped up on a pillow, listening to Tropical Fuck Storm then You Am I, King Gizzard, Magic Dirt, then someone else playing something else. The hum and patter of voices and guitars soaked through the wall. He slept there, on the floor. What else was there to do but close your eyes and wish it all away?

For almost three weeks, I held what was once a life with a beating heart in my body. During my lunch break one day, I was told by an Irish woman with an ultrasound machine that my baby no longer had a heartbeat. 'I'll be back in a moment, love,' she'd said, as I lay there in the orange glow, shielding my eyes from the screen. A second opinion was sought. But it only confirmed the first. For the next three weeks, in an apricot-coloured nook outside an obstetrician's office at Sydney's Prince Alfred Hospital, I existed in a liminal zone: I was neither pregnant, like those around me, nor not. I sat for hours, waiting for someone to tell me about a pill or procedure, or to remind me there was no heartbeat. And I was constantly hungry, had been for months; just over three to be exact, since I'd learned I was pregnant. For the first time in my life, I had a constant growl of want deep within me, and I satisfied every craving.

When I visited the hospital emergency room for the final time, it was just past 8am on a Monday and the hunger hadn't gone — it was my only constant companion. I had asked Pat not to come. The miscarriage was our loss, but it was my burden, and I squashed the grief alone in the time it took a taxi to drive from the hospital to my office at Sydney's Circular Quay. I was at work by ten, quieter, pale. And I started to see fault in everything. What I ate became a source of obsession, and questions about taste and texture were replaced with bigger thoughts about whether I was leeching on the world.

Beneath it all, I was whispering: *what did I do to deserve this?*

The shared lunch ordered for staff who rarely left their desks was spread on the kitchen table like a perverse offering: a pile of couscous salad on one plate, creamy red cabbage coleslaw on another. Tangy shredded chicken, slow-cooked lamb, buckets of hummus, piles of flat breads — a Lebanese-inspired feast for staff who only picked at the food. Half went into the fridge, and then into the rubbish bin the next day. Wasted.

One evening, my dad called to say, 'You should come home.' For more than 17 years I had lived abroad: first Melbourne and then Sydney, with a few years in London between. Australia felt like my home as much as the country I was born in — New Zealand (Aotearoa) — did. But it's also the place that fuelled a rabid ambition, which crumbled that August afternoon as the light was dancing and Magic Dirt was playing. Australia had burned me out. Consumed by to-do lists and outrage, I didn't know to stop when my body needed me to, and so my body did the most callous thing it could — it gave up on our baby.

When my dad said *come home, you can work on the farm. You*

and Pat can rear calves, it's a good way to start, I didn't know what he was suggesting. All I knew was it would be something different. I spent the first few years of my life on a farm, but despite that and my family's connection to the land — my divorced parents have both farmed and come from land-owning farming families — I have long embraced the rural–urban stereotype: the city is a place of culture and innovation; rural communities are conservative and culpable for vast amounts of environmental degradation. I had run to the invigorating embrace of the city as soon as I was able.

Pat was also burned out. Bent, exhausted, and pissed off, he agreed to move away from Sydney, his friends, his life for the same reason I did. He needed a reason to stop the 70-hour weeks, the Thursday nights that ended with a beer in front of the computer at 10pm. The Friday nights that ended with him falling asleep on the couch at 8pm. When we first met, he was an open-faced boy in second-hand Levi's and a plaid cotton shirt who bounced up gutters and into the nearest open band-room door. Back then he was fuelled by something that we had since squashed in our pursuit of … what was it? Money? Security? Accolades?

So, we said yes. To what? Rearing calves was the offer, though we didn't really know what that meant. We told people we were leaving to go cattle farming. To be farmers, food producers. I pictured vegetables in a garden, meat from the farm in the freezer, a life guided by food from the land around me. But I was naive. What I should have asked then was, 'How do you get closer to your food when the entire food-production system is guided by global commodities?'

Cattle farming is just one small part of a complex commodity system that dominates the Western world's food supply, a consequence of both colonial history and the Green Revolution.[1] When food was scarce in the wake of two world wars and global hunger was spotlighted as a national security issue, the global community decided collective action was needed and the United States responded with a burst of innovative technology — specifically, seed modification.[2] This technology was perfected in the late 1950s and distributed across the world in the 1960s, and it resulted in the global embrace of higher-yielding crops driven by irrigation and newly invented synthetic agrochemicals. The Green Revolution allowed communities to grow more food, but, with devastating efficiency, the system also homogenised the field and our palate.[3]

What do you taste when you think of the global palate? I long thought my cooking was reflective of the tastes of Sydney — a city rich with spices and flavours from elsewhere made local by migration — but in truth the city is a totem of the global food system that started with targeted plant breeding. Across the world, micronutrient-rich endemic crops and vegetables were replaced with just four crops: wheat, maize corn, rice,[4] and soy.[5] By 2009, wheat was the dominant crop grown in the world and remains so still.[6] In Sydney, I could stop off at Woolworths on my way home from the office to pick up tomatoes, basmati rice, and saffron for a paella. Easy access props up this global food system, and food is a luxury, a salve.

What meaning do you ascribe to food? Is it a symbol of culture, community, jobs, justice, environment? Or is it just about taste? Food has meaning at Pat's and my table: it squashes for a moment the sadness that started with our loss and grew as I surveyed the horizon. Yet before it gets to the table, it is really only one thing — a tradeable commodity with a trade price set in advance and far away from the place where it is grown.[7] Beef slots into this system seamlessly.

At dinner, not long before we left Sydney, a close friend joked to her new date that we were running off to New Zealand to keep chickens and grow vegetables. The image my friend conjured is of a lifestyle — a few acres with some land, a cow, a pig, and a few chickens. I blushed hot and snapped back: 'It's a real farm.' We're moving to a 130-hectare cattle farm. A leased farm, not an inheritance. A business, not a family tradition. Big enough for it to be enough.

I'm going to be a cattle farmer.

I want to prove I'm not dropping out or quitting. The farm is just another challenge. I want to prove to all those who are amused, and assume we'll be back before the year is out, that I am strong enough to succeed.

I spent years working my way up in the media industry, through magazine publishing, then the Australian Broadcasting Corporation, and finally BuzzFeed. From assistant roles to editorships and executive positions, I thought of myself as a quiet operative in the glass-ceiling war, but really I was outraged by everything and nothing. The media game was exhausting and bankrupt. I was drained and muted. So, we said yes to becoming farmers in December, and we moved in February.

When people ask why we did it, I say with all earnestness that I wanted to be a farmer so I could produce food — the one thing that had filled me up, brought me pleasure, made me whole again after the miscarriage. I wanted to be close to the root of it all. But equally, I did it to prove other people wrong. My dad, who suggested it as a joke and thought we wouldn't last; my friend, who thought it was a lifestyle choice; my former colleagues, who thought I was giving up; and everyone else who told me there's no such thing as a meat-eating environmentalist — although that part came later.

A few months before I fell pregnant, I flew to New York for a work trip. BuzzFeed is headquartered on East 18th Street in Manhattan, and it had been decided that a few face-to-face meetings might be useful, as there's a limit to what can be communicated on internal chat platforms. Hungover with jet lag, I sat one morning at a table with the then editor-in-chief, Ben Smith, and chatted about our editorial priorities for the year ahead. I pitched an idea to cover climate change in a new way. The Standing Rock protests were happening 2,600 kilometres west in South Dakota. Thousands of people protesting the Dakota Access Pipeline had been demonstrating at the Standing Rock Sioux Reservation for almost a year, and they'd come from all over the world to do it. BuzzFeed had sent reporters to cover the protests. *What if we expanded that beat to Australia?* I said. *Cover climate change and environmental change through land rights? No one cares,* Ben replied. And he was right. News about environmental degradation was hard to sell no matter the packaging. Standing Rock disbanded, people kept driving their cars to work, flying away to holidays, consuming, eating, consuming more. Life went on. So did climate change.

I walked downtown to meet a friend at Dudley's, an Australian bistro on Orchard Street that served delicious avocado toast. He was over from Paris for work, one of dozens of flights he'd take that year. We had lived together in Melbourne in 2006, and then I had left for London and he'd moved north to Fitzroy. That same year, a few people in various positions of power had started speaking passionately and publicly about the importance and danger of global warming. But neither of us paid much attention; we were 25 and our lives were opening up in front of us. I remember that Al Gore — the former United States vice president turned one-time presidential hopeful — released a film called *An Inconvenient Truth*. People watched it and were shocked and concerned by Gore's claim that we were in a 'planetary emergency', but for the most part we all went on with our daily lives as we had before. I flew to and from London twice and thought nothing of the air miles or carbon footprint.

This was also the year the United Nations' Food and Agriculture Organisation (FAO) issued a report titled Livestock's Long Shadow proclaiming that rearing cattle produces more greenhouse gases than driving cars.[8] I missed it in 2006; it didn't get quite the same attention as An Inconvenient Truth. Two Academy Awards and US$50 million at the global box office did not follow. Instead, a raft of bad press and the issuing of a 'technical note' were to come. But at the time of release, the report's findings were as grave and urgent as Gore's. Senior FAO official Henning Steinfeld said, 'Livestock are one of the most significant contributors to today's most serious environmental problems.'[9]

The report's researchers figured that the livestock sector (the farming of cattle, sheep, pigs, and chicken) generated *more* greenhouse gas (GHG) emissions when measured in carbon dioxide (CO_2) equivalent than transport. They put the sector's total GHG contribution at 18 per cent of the world total, and the media swooped. The fossil fuels industry was no longer the only baddie in the climate change story. I remember that part — the appearance of cows in headlines where once billowing smokestacks had loomed.

Steinfeld had called for 'urgent action' but what he got was the start of an ideological battle and reductive media coverage because colleagues disagreed with the FOA team's research methods: they'd counted the cars but not the manufacturing or the road construction, for example.[10] The FOA's research was questioned, and in the years that followed it revised its calculations and issued updates that put the GHG figure for livestock farming at around 14 per cent. But the media now had a simplified version of complicated climate research: *cows are worse than cars.*

It didn't occur to me then or later that I should connect this 14 per cent to the pastime that was lifting me out of heartache. Cooking and climate change seemed like odd bedfellows, so I spent long, slow days in the kitchen, combining chilli, turmeric, cumin, coriander, and paprika with lamb, yogurt, and tomato: the start of a biryani. A pungent mix of celery, carrot, onion, garlic, mushrooms, and herbs with beef chunks and stock would reduce down to a base on which to layer fluffy, light potato, to make a pie.

I made hearty meals to soothe my searing insides. But those meals were a point on the emissions scale. That became clear to me in 2018, when I made the decision to leave Australia to return to my other home, New Zealand. I convinced my Australian husband it would be an adventure. But the decision to quit city life to become cattle farmers dropped us amid a cluster of arguments about food and farming and its role in causing and combating climate change and the degradation of land, fresh water, air — the resources that make up our environment.

At first, I tried to win every argument I stumbled into: veganism versus meat eaters; clean food versus food tech; regenerative versus conventional farming; the North's growth versus the South's ecological scars. The battles consumed me, and very near destroyed Pat and me.

And then I stopped fighting the status quo and immersed myself in creating something new and small. This book is about how a plot of leased land in the middle of New Zealand helped me to untether from the old world, physically and emotionally. It's a book about food, farming, and climate change — our collective challenge — and how they have been co-opted by marketers and lobbyists dictating solutions. I've tried to answer every question I've heard asked: is beef bad for the environment? Is veganism the answer? Is local always better? Can food technology save us all? Is climate change *my* problem? Why is food tearing my family apart?

But it's also a book about sacrifice, because I do not see a future without us all giving up something in order to create something new, something better. That shouldn't be a cause for fear. The human impulse to break and remake afresh can result in a thing of startling beauty. This I know from looking at that sculpture for so long, from cooking food, from learning to read the land, and from the struggle to become a cattle-farming, meat-eating environmentalist.

I reached for the sky, refusing to fall beneath the surface.

1.

Big bad beef

'How do you catch a calf?' I whisper to Pat as he backs the truck and trailer up to the pen gate. He shrugs. 'Can't be that hard, they're little.' The calf shed is a ramshackle cluster of small pens with white plastic sheeting covering the gates, shielding the small animals curled into balls from the wind battering the walls.

The first is hoisted off the floor half asleep and into the trailer, then the rest scatter, running ribbons around us. We back them into corners, grabbing for an ear or tail. They're stronger than they look. At the trailer, Pat encircles his arms around the legs of each animal and lifts it in.

With stains of putrid milky poo spread across my jeans, I stand on the edge of the trailer, counting our new acquisitions. Twenty-two calves — a motley selection of crossbreeds from Friesian and Jersey cows and Hereford and Angus bulls. Some are well marked with white faces, white socks, and a white tail tip. Others are six-point Friesian bulls with a black-and-white patchwork coat and a white heart on their forehead. A few have a red mottled coat and splotches of colour over their face. We buy them because we don't know not to. Turns out other farmers don't want those ones. But all are destined for the beef trade eventually.

I've started my first day as a farmer dressed in old jeans, a hand-me-down vest, a black windbreaker, a new Patagonia cap I bought in Sydney, and Redband gumboots. I'm doing my best to look like a New Zealand farmer. I feel like a fraud.

Two weeks ago, I phoned a woman who owns a dairy farm, hoping she'd be kind and understand that I didn't know what I was doing. 'I want to buy some calves,' I stammered. 'I can't talk right now, I'm driving,' she replied abruptly once I'd finished my introduction. 'Call Chris. I'll message you his number.' Chris, I learned, was the farm manager, and he agreed to meet us and introduce us to the calf rearer.

We were due at the dairy farm, located 30 minutes from our plot of land, at 9.30am. We didn't arrive until almost ten o'clock. The delay was my fault. I had scribbled the address on an envelope and hadn't asked for directions. My habit has always been to figure it out on the fly: the sign of workplace competence. The night before I had googled the road, and in the morning, I directed Pat to the other end, some 20 kilometres in the wrong direction. He drove silently, cursing my carelessness. I sat in the passenger seat, staring at the map on my phone and blinking back tears, cursing him for being so critical. I didn't think I needed to change. *Project confidence and just figure it out* — that had been the way to get ahead in my previous career. It didn't occur to me that I was embarrassing my husband and business partner — who was particular about directions, expenditure and budgets, and the like — and slowly revealing to him something I'd kept away from our home life until now: I was a bit of a blaggard.

Our lives in Sydney had circled furiously in close proximity to each other, but barely overlapped. The dense, unflattering core of each of us had remained intact. Now, we were working, learning, living together; one another's only friend and confidant. Slamming against each other as our new life tried to merge us into one.

The air is hard in this part of the world at this time of the year. It dries my skin, creating cracks between my fingers where dirt clusters. Each morning the sky lights up a startling, brittle blue. The scene out my bedroom window is a vista from a settler painting: strict outlines, crystalline sky, mountain ranges outlined by heavy brown or dark-blue lines above cleared farmland. In one corner the view is interrupted by a hardened burnt tree stump; beyond is a nikau palm standing obstinate against the forces of progress. 38.6843 degrees south; 176.0704 degrees east; 360 metres above sea level. Taupo. The Central Plateau. Volcano country. Our new home.

For five weeks in autumn, we drive the rural backroads collecting calves from dairy farms. As we near the end of our first season rearing calves, the story that farming, and specifically cattle farming, is destroying the planet suddenly lands on the front pages of newspapers around the world. It's 2006 all over again. 'The Environmental Cost of Eating Meat Is Turning People to Veganism' (*BuzzFeed*); 'If Everyone Ate Beans Instead of Beef' (*The Atlantic*); 'Eat Beans, Not Beef to Save the Planet' (*Politico*); 'Avoiding Meat and Dairy is "Single Biggest Way" to Reduce Your Impact on Earth' (*The Guardian*) — the list goes on, and I pay attention. And I start to ask questions.

Are the headlines right? My dad and his friends treat my inquiries as partisan accusations. 'Just bloody green city lefties blaming us for things,' they say.

We keep working: months of seven-day weeks, then a few weeks off, then the calf pick-ups start again for another ten weeks in spring. Each dairy farm sells us between four and 40 calves a day. The animals are mere days old when we transport them to the 130-hectare lease property we share with my father. We unload them into an old hay barn that Pat has refurbished by ripping out one of the corrugated iron walls to create ventilation, and building small pens floored with wood chips. Before each little animal has an ear tag clipped onto it, its navel is sprayed with iodine to ward off infection and it receives a shot of vaccine under the neck skin to prevent sudden death from a range of diseases. We learn to do all this by watching YouTube videos.

I am constantly anxious.

It's a wet morning when we pull up to the converted deer shed that houses young calves on one of the dairy farms that Chris, the farm manager we met on our first day, manages. The steep farm track leading to the shed is washed out and muddy.

'It looks too steep,' I mutter in Pat's direction.

'It's not.'

'It looks too muddy.'

'No. Just direct me.'

I'm trying desperately to control *something* — everything in this new life feels so chaotic.

'I'm not an idiot,' Pat mumbles as he pulls the hand drive to park, having made the climb easily. 'You need to trust me.' I turn away from that comment and walk inside to find black-and-white calves curled into warm balls, sleeping through the morning. A South African woman is dressed in weatherproof overalls and a wet jacket, ready to help steer 20 calves through the shed's walkways and up our trailer ramp onto the hay. In Afrikaans, she asks the eldest of her three barefoot blonde children, who are climbing the rails, to help us with the gates. It is a process we've done before, but this time I don't check the sex of each calf as they're loaded into the trailer. Among the Friesian bulls are two small heifers — female calves.

Their presence becomes a running joke and ammunition for my dad, who likes to point out that our knowledge is so limited we don't know a male from a female cattle beast. I ask him how to tell the difference when we're in the yards drafting our first mob one clear morning, separating them into groups of male and female. His eyes crinkle and a deep burst of guttural laughter explodes. 'One has a pizzle!'

'A what?' I ask.

'A pizzle,' he cries again while jabbing his mustering stick — a piece of PVC pipe — at the underside of a steer's belly.

I learn later that a dried pizzle is also a 'tasty, healthy treat' for dogs. Castrated bull penis makes for a good chewy, apparently.

That same day I watch him cut off a two-year-old cattle beast's small, pointed horns with a pair of horn tippers — a contraption that looks a lot like tree-pruning shears — causing jets of blood to jump and twitch over the animal and the yards. He is about to sell the animal, and the horns must be removed before trucking it.

I return to the yards alone the next day. No one sees me with a bucket of warm soapy water, futilely scrubbing the wooden railings clean of blood. I vow to look elsewhere for advice.

I should have taken my own instruction more urgently. My father had volunteered to be our mentor and farm adviser. He had told us, before we left Sydney, that rearing calves would be profitable, and we wanted to be in the food and farming business — not just to while away a few years, and all our savings, living a romanticised version of life on the land for the sake of *lifestyle*.

I believed that everyone was honest. That my dad, the dairy manager, the stock agents, and the farm store manager knew what they were doing and were guiding us in the right direction — not just making money from our labour.

We'd purchased a new trailer and retrofitted it with plywood sides and a half roof to shield the young calves from the wind and rain on the drive back to our farm. We'd rebuilt our hay shed with the help of an old fencer. The bill arrived: NZ$2,300. We purchased hay from a neighbouring farmer: $50 a bale. A fertiliser bill arrived: we were to pay half of the $6,000 and we'd barely spent a day on the farm. We were dishing out cash quickly. Five thousand dollars for a new stock trailer; $200 for each calf; $105 for each 20-kilogram bag of milk powder, and we were buying it by the tonne; another $30 for each bag of protein-rich calf pellets.

And then there were the favours.

'I've got hold of him,' my dad tells us as we're sitting around his dining table one evening early in the first calf season. I have a glass of red wine in front of me; Pat stands and goes to the kitchen to collect the bottle, refilling his glass. Bracing for what comes next.

'But no word yet on when we can go get those calves. It'll happen though, don't worry.' This is the farm advice we're paying him for. Calls to mates about deals that never eventuate. Dad's mobile phone rings, and he answers with a blustering ''ello'. Leaning forward over the table he settles into the conversation. Pat looks at me, his eyes watery, expression pained.

'Tell your dad we don't want them,' he hisses. 'I'm not rearing 20 calves for free for this guy.'

'I know, but what else can we do? I don't know who else to call,' I plead. We've devised a business plan, of sorts, that requires we buy no less than 250 calves a season. 'If we don't buy his calves, how are we going to source enough?'

Every day there's some new deal we need to know about. In return for the use of 30 of the 130 hectares of farmland, plus tools, equipment, and a house, we are to pay my dad $3,200 a month and buy and rear 20 calves for him, which he'll sell two years later for $1,100 each. His mate wants us to do another 20 calves for him, free of charge, in return for a steady supply from the large corporate farm he manages. We are told this is how people do business in farming: a favour here, a favour there. It comes back around. You make us money; we'll make you money.

But they were laughing at us. To them, we're just fresh faces from the city with money to burn. That money, for us, is days, weeks, years of careful saving and hard, draining labour. I want to believe so much that these people in our lives are good. But in farming, there is always someone at the gate with their hand out, waiting for payment. And I thought I always had to pay.

Pat ended up paying more.

As the months pass, and autumn turns to winter, my life on the farm starts to form a pattern. Early, around 5.30am, I awaken, drink a mug of coffee, eat a piece of toast, and kit up in layers of wool — leggings, thick socks, a jumper, then a down vest and a windbreaker. My walk to work is short: out the back door and gate, down a short path under towering pines and into the farm compound where our calf shed is located. The perfume of the rotting pine needles on the track, the dew laying heavily on the grass, and a hint of wood smoke give me a rush of energy each morning.

The next three hours are consumed by monotonous labour: I mix hundreds of litres of warm water with bags of milk powder, lug feeding containers with rubber teats over fences and into the shed, wrestle week-old calves onto the teats, scoop up mounds of calf manure in the pens, scrub down the pen rails and water troughs with disinfectant, and replenish the containers and troughs with fresh water and hay. Unseen bugs are the calves' enemy, and my work is to eradicate any trace of them. Feed, clean, feed, clean.

This is my introduction to farming.

In the late morning, I start to cook. A heavy labour-induced hunger sits in my bones; I'm ravenous. Toast, followed by scones, then eggs for lunch, more snacks in the afternoon — fruit and biscuits — then dinner, usually meat.

I turn the oven to 180 degrees and start to crumple butter into flour. Add a pinch of salt and teaspoon of sugar, then a teaspoon of baking powder. Rubbing the butter between my fingertips, crumbs start to form, and I dribble a little milk into the mixture and start cutting it against the side of the bowl with a knife to bind. Milk, cut. Milk, cut. The scone dough comes together as a wet mess. Finally, I crumble in feta cheese and chopped spinach, and drop the lump onto a lined baking tray. I press the mess flat, fold in two and repeat. Then I form it into a round, scour the top with a knife, and slide it into the hot oven. Twenty-five minutes. I count. I wait. The smell of a crisping scone fills my head like a smooth tonic.

Next, I slice an onion, paper thin, into pungent ribbons. Peel and slice four garlic cloves and a thumb length of ginger root. Into a cast-iron pan I drop hunks of fibrous chuck steak. I leave it to brown on each side and then scoop it out into a bowl to let the chocolatey juices amass in the base. The pan is wiped clean. A dollop of butter and then the onion. Once the onion is clear and caramelising, I add the garlic and ginger. A heady aroma wafts from this simple mix; I lean over and inhale deeply. Soon I return the beef to the pan and start scooping three tablespoons of my berbere spice mix: for every tablespoon of paprika a combination of dry-roasted cardamom, cloves, coriander and fenugreek seeds, black peppercorns, and allspice is added. Plus, chilli and salt.

As the mix goes dry over the heat, I add half a can of tomatoes and enough stock to submerge the meat. Cover and turn the heat down low to simmer for the hours ahead. I slather my savoury scone with butter and eat it in four mouthfuls. Pat's portion is long gone. Then it's back to work. The routine repeated. But when we return, a small luxury awaits. Slow-cooked berbere beef. A jammy, rich meal that fills me up.

By October, the snow is still lying heavy on the mountain range at the base of the lake just south of the farm. My hands and toes bear the scars of chilblains, a reminder of the early mornings feeding warm milk to cold calves in temperatures that never get above zero degrees Celsius. The air is heavy and wet, and it brings earthworms to the surface of the mud-covered yards.

We've signed a contract with an agent to get 150 of our calves to 100 kilograms before November, so that another farmer can fatten them on late spring grass and hit their own processing deadline to kill the animals before they reach 24 months.

We push. The small, cold animals are fed twice daily: first, a warm milk mix made from powder that costs more than the calves themselves, followed by meal made from a concoction of local and imported grains and vitamins, followed by hay, and finally grass. But they look *hard*, which is to say they don't have the round rump and thick neck typical of a calf that stays on the cow.

I'm in the grind of labour, and there's little evidence of my grand vision of becoming a food producer. Around me are veterans of farming pushing us to take on more calves, to sell more, to fatten more. For so long, the mission of farming here has been to bend nature to man's will to increase yield and profit, not to produce good food within the limits nature has set out. These older men see potential profit in our endeavours, not the drudgery of mud, brittle frost, chilblains, bruises, and sadness. I am heartsick because in this push, calves have died. Some of those little Friesian bulls we chased around the shed have come to the farm sickly and haven't lived beyond ten days.

This is my farming reality.

One small bull calf has stopped drinking milk, and I've left it lying limp on the hay-covered floor of the shed. Exhausted and bone weary, I drop next to it and cradle its head in my lap, slowly dripping a stream of warm milk into its mouth to encourage it to suckle. It latches on to the pink rubber teat and lethargically pumps for milk, but it's not drinking enough. The calf has scours and is losing vital fluids through diarrhoea. It's common when the calves switch from the colostrum-rich cow's milk to a milk replacement. Their little digestive systems take time to adjust to the changed diet, and the stress of moving farms and bad weather also have an impact.

This is not the first animal I've tried to nurse back to good health this season. When the vet arrives to assess the calf, he says it's cruel to keep it alive. The technician accompanying him looks at me sitting on the hay-covered pen floor dressed in manure-stained waterproof overalls and a wet-weather jacket and says, 'I've never seen a calf rearer do that.'

She means it as a compliment for the care I'm taking, but I feel like a failure.

Our animals shouldn't be dying.

My business, and the wider New Zealand beef and dairy business I'm part of, is in part responsible for this calf's suffering. Nature intends for this calf to stay with the cow for the first months of its life; business intends for the mother cow to be in the milking shed a day after calving, in order to supply milk, which becomes powder and other milk-solid products, which are traded into foreign markets for six, seven, eight dollars a kilogram. All adding up to a dairy export industry worth NZ$18 billion in revenue.

That same business model expects the cow's progeny to be reared by people like us, fattened by farmers with more land, killed after 18 or 24 months of life, and exported as premium beef to Japan, Hong Kong, or Taiwan, or as hamburger meat for the United States or China.

More than 450,000 tonnes of beef meat are exported from New Zealand each year, to North America and Asia,[1] mostly. Australia, one of the world's largest exporters, sends more than double[2] that abroad.[3] The China market alone is worth NZ$1.8 billion to New Zealand, the United States NZ$1.3 billion.[4] And it's possible because both countries have historically played fairly with global trade partners. Free trade agreements get meat products over borders, with few tariffs, making beef farming a profitable business — if you're only in business.

On an oppressive grey spring evening back on our farm, our vet euthanises a dehydrated, waning Angus Friesian bull calf with a bolt-gun shot to the head. Another two died while they slept the day before. His parting comment is an aphorism intended to buoy my spirits: 'If you've got livestock, you've got dead stock!'

I wanted to get close to my food source for my own health, initially, but the degree of closeness has started to put me off my food and my new vocation. I walk home from the calf shed, back under the pine stand that now has a funky smell of rot and decay. For a moment I stop, and slump against a tree stump, surrounded by tall grass. A heave of sadness bursts from deep within me. 'What have I done?' I gasp. The grass tickles the side of my cheek as I lean my head against the bark, wanting it to absorb something from me, to absolve me. I am not so hardened as to feel nothing at the death of a sick animal.

Pat stays back to bury the calf. Somehow this has become his job. Disposal. And he's started to disassociate; I can hear it in the few words he shares at the end of each day.

The animal now represents lost money. It's a coin to be buried.

'We've got to keep going,' he says. 'I've spent too much money to give up now.'

We can't keep going like this, following advice that seems so outdated. But we have to keep going.

Ask how we got to this point, and most will say 'the government told us to do it'. The government wanted productive farmland to increase export revenue, and that required the clearing of nature. When my dad started farming back in the 1970s, Britain was about to join the European Union, ending a longstanding colonial trade relationship that positioned New Zealand as Britain's 'farm in the South Pacific'. During this period, the government kept farmers afloat through subsidy-grant-assistance payments, and hard men in black singlets were encouraged to 'break the land' and make it productive.[5] They drained wetlands and turned it into pasture, back-burned native shrubs to create hill-country grazing, felled trees, and sculpted the land. Top-dressing aeroplanes would fly low over newly cleared farmland, dropping tonnes of synthetic fertilisers containing combinations of nitrogen, phosphorus, calcium, sulphur, and other minerals.

For centuries farmers have used natural fertilisers like bird droppings, manure, seaweed, and the remnants of decomposed animals to add nutrients back into the soil to help crops and grasses grow. Plants need nutrients to thrive, and nitrogen and phosphorus are chief among them. The world cannot be kept in food without fertiliser. But manure was replaced with synthetic fertiliser in the twentieth century with the invention of the Haber-Bosch process: a breathtaking innovation that saw manufactured materials used to make explosives for World War I repurposed as synthetic nitrogen. It was the start of the fertiliser boom. Farmers embraced the theory that applying more fertiliser was the fastest and easiest way to increase yields; suddenly they could grow and sell more food.

But if the combination of synthetic nitrogen fertiliser and nitrogen excreted by animals via urine and manure exceeds a plant's needs, it becomes a problem. My dad didn't know it at the time, but all those tonnes dropped from the air were the start of something damaging. And he didn't stop to ask if it was a problem later, because on the farm that fertiliser grew more feed, and more feed meant more animals and more profit.

Through the porous soils the nitrates leached, for decade after decade, until suddenly the rivers and bore wells started to change.[6]

The water became ripe for algal blooms and other subtle ecological changes. Drinking-water wells in agricultural areas showed high levels of nitrate-nitrogen, making the water unsafe to drink, especially for bottle-fed babies. 'The government told us to do it,' my dad says. 'Back then, using fertiliser was encouraged.'

Dad stopped farming for a while in 1989. The farm of my childhood was sold. New Zealand's Labour government had four years earlier ushered in an era of reforms that cemented free-market neoliberalism at the centre of the New Zealand economy, plugging its agriculture into the global trading system. Farmers lost up to 30 per cent of their annual income overnight as the subsidies disappeared, and some, like my dad, didn't survive the change.[7] Sometimes it feels like we're still in the 1980s — a time when hard men worked the land and complained about change.

Those that survived, flourished. They started to push the land even harder. Intensification tore through the countryside as farms once used to farm, sheep, cattle, and crops, often together in rotation, became vastly more profitable dairy farms. But to farm more cows per hectare, farmers leant on three main tools: irrigation for constant water, energy to drive the irrigators, and fertiliser to grow the grass. Between 2002 and 2012 the amount of New Zealand land under irrigation almost doubled, from 384,000 to 735,000 hectares.[8] Fertiliser use shows a similar pattern: by 2016 the overall average use of nitrogen for dairy farms was 126 kilograms per hectare, up from 40 kilograms per hectare in the late 1990s.[9]

Productivity became a religion, and it was luminous and blinding. The polluted water, the changing climate, the vast divide opening up between people who produce food and those who consume it — none of these is a reason to change the way things have been done productively around here for decades, they all say.

I believed these teachings for a while and didn't ask if there was another way. Didn't pause when something felt wrong, as it so often did. The confidence I had felt in Sydney had disappeared at the same rate as our savings. I'd believed then that we were skilled enough to be good farmers, that my inner-city ambitions to care for the environment and produce food that was affordable and nutritious with care and consideration would survive our move to the country. But I was out of my depth. Fuelled by a fear of failure I focused on profit above all else, and as a result I created a little industrialised system on our farm to slot neatly into the big industrialised system that governs New Zealand beef and dairy. Within months of arriving back in New Zealand, I was essentially factory farming, pushing products — animals — through a production line as quickly and efficiently as possible. And lining everyone's pockets, except for ours. There was little care in what we were doing; we were just trying to survive. That's how the trouble started.

'Do you talk to your mates about climate change?' I ask my dad and uncle one afternoon as we're sitting at the picnic table in front of the house. The warm sun tickles our bare arms.

'Yeah,' my uncle says and then chuckles.

'Is it a joke,' I ask, 'that I worry about it?'

'Well yeah, you lot say it's a bit more serious than we do,' he says. 'It *is* warming, but most of us won't say it's because of farming.'

'What do you call us? Boomers?' my dad asks, cutting in.

'Yeah, boomers,' I say. And then I tell him what's been bugging me: that his generation seems to think climate change is a joke at best, or worse, an issue cooked up by leftie urbanites. By people like me.

I ask Dad if he thinks we should stop buying young bulls and fattening them for the processing beef trade — in other words, making burgers for the fast-food industry — because all it does is put more cattle on the land for the sake of fatty burger beef. 'If we didn't do it someone else would,' he replies.

That's been the thinking here in New Zealand and in Australia for the last 50 years, he says. *Better us than someone else. I may as well make the dollars rather than him.* Better we shoulder the environmental damage? Not exactly. Because my dad and his lot suspect the damage isn't all that bad. That this climate change conversation might be a bit of a fad.

'Go back to the 1800s, how many buffalo roamed the plains?' he asks me with all seriousness. He's collected this argument along the way. Heard it once and stuck with it because it suits his world view. There's long been large herds of ruminant animals burping and farting their way across the land, so why are we being punished for it now?

A small part of me wants to be accepted by my new community, these farmers. But I also want answers. I can see the land cracking; so can Pat. His quick laugh and the playful crinkles that clustered and jumped at the corner of his eyes as we talked are gone. Replaced with long silences and inert eyes. In darker moments, Pat turns to me and mutters, 'Just stop trying … they won't change.'

I can't. To stop trying is to admit that we've made a mistake coming here. That farming is unchangeable, and that profit is the only ambition. I don't have it in me to say this aloud. Not yet.

It's dry. The pasture plants are topped with clusters of seeds, and the wind sends them aloft to hang heavy in the air. Some float into the faces of our older calves as they graze over the steep hill face of the paddocks behind the house. The seeds lodge under their eyelids, causing their eyes to weep. The seeds burrow in, like a small creature searching for a haven. And the weeping continues. If I intervene early and apply an ointment to flush out the seeds and treat the irritation, the weeping will stop, but if I miss it, miss the signs, soon a bulbous lump will form on the eyeball, and with the threat of blindness present, our vet will return to operate. The eye is stitched closed until it's healed. A vet bill arrives for $320, more than the value of the animal, and summer continues.

It doesn't rain properly until May.

'Don't worry,' they all say. 'It'll rain, it always does.' But the farm is overstocked. We are moving mobs daily to chase blades of grass. Profit-chasing and climate change do not marry well. A sense of panic has descended.

One morning, Pat and I discover a three-metre-tall mound of palm-kernel extract has been dumped in the middle of one of the holding paddocks leading into the cattle yards.

'What the fuck is that?' Pat says, shaking his head.

Palm kernel. An imported gritty meal made from the by-product of palm oil production in Southeast Asia. Short on feed for the dairy heifers my dad is being paid to graze on the farm, he's brought in feed sourced from a wasteful industry responsible for deforestation in Indonesia and Malaysia. But it's also considered cattle food that contains some protein and a little bit of energy, and it's cheap. And we have no grass.

A few days later, Pat discovers a full palm-kernel trailer fishtailed across the farm driveway, halfway up one of the steep hills. It's blocking the drive. My dad has tried to pull the heavy trailer with the four-wheel quad bike. It isn't powerful enough — the deep skid marks carved into the drive are proof of that. He lost control and shunted the trailer sideways to prevent it sliding dangerously down the hill with him and bike attached. And then he just left it there and went home. For lunch.

'So, what, I'm expected to tow it out with the tractor?' Pat asks me.

'I don't know...' I reply.

I don't know anything anymore.

'I'm sick of this shit!' my husband yells across the paddock. A trailer full of environmentally corrosive feed supplement stuck across the driveway was the thing that broke us. Pat retreats to our bedroom. He stops speaking to my father, who becomes a symbol, fairly or not, for all that is wrong with the farming sector.

I no longer trust that anyone around me knows what is happening to us, to the land. *I've made a big mistake coming home*, I think.

'You should go see Bill,' my uncle says. 'Then you'll understand.'

He is trying to be kind and help me to understand that not all conventional animal farming mirrors what we do in our converted hay shed. It's not all intensive.

'Conventional' is the word used to describe most New Zealand farms. In simple terms, it's a style of farming that relies a little or a lot on chemical inputs, like fertiliser and weed, and pest control to make grass and crops grow — conventional farming is a by-product of the Green Revolution's technological advances. One alternative is organic farming, or systems that follow the principles of permaculture or agroecology. Bill's methods are conventional but different, my uncle assures me. He's trying to give me hope.

Does that make them any less damaging for the environment, I wonder?

More than a year passes before I drive south to visit Bill and Jennifer at their immaculate cottage in Mataroa, 18 minutes west of Taihape in the middle of New Zealand's North Island (Te Ika-a-Māui). Clusters of dozing daffodils flank their driveway, and beyond the road lies Paengaroa Reserve, a Department of Conservation–protected island of native vegetation that shadows the Hautapu river. 'It's a work in progress,' Bill remarks, gesturing to the flowers as I step out of the truck to say hello. 'Jennifer has been trying to map them when they're out,' adding that the plan is to create a wall of yellow. No gaps between blooms — a clear colour line between the controlled confines of home and the deep-green depths of the bush beyond.

Inside, and straight to business. I'm presented with a map of the farm. A thousand acres: 600 purchased in 1986, another 400 in 1990. 'That's in bush,' Bill says, pointing to the paddock named '40 Acres' on the small satellite map, and continues: 'That's in native.'

'How long have you been farming this block?' I ask.

'I think it's close to 45 years.' Each year seems woven into the weather lines on his red, vein-flecked cheeks. He was born and raised in this valley, affectionately known as Coogan-ville. Cousins farm either side of the road leading up to his farm. Bill's grandfather arrived in 1886.

'I thought I might hang out to 50 [years],' Bill says before elaborating that he's becoming disillusioned.

'With what?' I ask.

'Oh, a lot of these regulations and rules that are coming in. All this greenie stuff—' He's talking about the government-led freshwater and climate change reforms that require farmers to fence off waterways and plant riparian barriers to help mitigate water pollution, and retire steep or unproductive farmland to make way for tree plantations that will, hypothetically, act as carbon sinks in the future. Bill drifts off, then eventually makes clear that he reckons the best regulator of farmland — in fact, the best protector of land — is Mother Nature. The trick is to listen to her. Perhaps that's where I've been going wrong. She is the guide, not them; not the regulators or the old boys.

Hours later I'm at the top of a steep, winding single road, trying to stay upright. A rough northerly wind is howling down the Mataroa Valley from the foot of Mount Ruapehu in the central North Island and it's barrelling straight up the cliff face to where Bill and I stand. I follow him to the cliff and look over the edge. Wind-blown tears run from the crinkled corners of Bill's pale-blue eyes. 'It used to scare me a little,' says Bill, about the steepness. 'But it's dead easy now.'

As he gestures towards a stand of scrubby bush clinging to the face, he tells me about the days when the sky glowed orange from the back-burning of manuka tea-tree. My eyebrow shoots up and he notices. 'There're no hillside fires now,' he chuckles. 'But back then the government encouraged it.' Echoing my dad. Now the government and people like me are asking him, them, to change again, to follow environmental regulations that are intended to curb emissions, yet this time … he doesn't think he should listen.

When I ask Bill how his farm will cope with the changing climate, he says it will be all right as long as there's water and trees for shade. When I ask how he knows if he's improving or degrading the property, he spouts aphorisms about learning to read the land. 'Mother Nature has a terrific way of healing,' says Bill optimistically, making reference to a track being bulldozed through farmland only to be overgrown with pasture years later. It's nature, of sorts, at work. 'I don't think a lot of people realise that, especially policymakers.'

On Bill's farm, there is no environmental problem. The land is productive; tree stands have been left in place, working to cool the land and suck up carbon; the animals are well fed and free to roam. It's old-fashioned farming, Bill likes to say. Controlling nature but listening to her, nonetheless. He can say the problem doesn't lie with him, but I think to do so in a small, tight-knit community like his is to give permission for his neighbour, and *his* neighbour, and so on, to say the same. The decisions these landowning men make in the very near future will determine whether we'll be truly able to solve our environmental problems. And it *is* largely men. A research paper released in 2020 by two well-regarded agriculture universities surveyed 500 farmers to gauge the pros and cons of engaging in bio-diversity programs on-farm. Ninety per cent of those surveyed were men over 45 years of age. The researchers called the home number listed on a farming registry and asked to speak with whoever 'makes most of the decisions', and so the men spoke.[10]

What power do I have? Bill has 1,000 acres; I have much less than that. Bill owns his property. Now worth around NZ\$3.5 million. I pay rent. He doesn't think he needs to change; I feel in my heart we must. Which of us is right?

In 2019, the International Panel for Climate Change (IPCC) researchers issued the second in a series of reports suggesting strategies to keep global warming below two degrees, a task only achievable if all heavy polluting sectors, including agriculture, reduce their collective greenhouse gas emissions. Agriculture, forestry, and other land use accounted for 23 per cent of the total net man-made emissions of greenhouse gases in 2016–17, and 44 per cent of total methane emissions. The report stated that a shift to more sustainable food production offered the best chance to tackle climate change.[11]

It's a global problem, and a global solution is required, which pulled the IPCC working group towards one salient point, articulated here by its co-chair Debra Roberts: 'Balanced diets featuring plant-based foods, such as coarse grains, legume, fruits, and vegetables, and animal-sourced food produced sustainably in low green gas emission systems, present major opportunities for adaptation to and limiting climate change.'[12] My eyes landed on 'food produced sustainably', but fixated on 'plant-based foods'. Two points with the same goal — food on the table — but two points with very different ramifications.

Back in 2006, when the United Nations' FAO released its report *Livestock's Long Shadow*, Australia and New Zealand were consuming around 26[13] and 28[14] kilograms per capita of red meat respectively — we are some of the biggest meat eaters in the world. We're far behind people living in the United States, though, who eat an average of 26.3 kilograms of *just beef* per capita each year, second in the world only to Argentina.[15] Until recently the United States was both the biggest importer and producer of beef. To eat beef is to be American. It's a symbol of prosperity.

Sophie Egan, an author and program director at the Culinary Institute of America, suggests that the consumption of beef has been embedded in the American myth of manifest destiny; it's a philosophy fixated on territorial expansion via a conjoining of democracy and capitalism, and it took root in Australia and New Zealand too, albeit to a lesser degree. The dream, she says, is 'having access to that meat locker'.[16] But it's what's in that locker that's relevant to me. It's a lot of processed meat. That's where my little Friesian bulls are headed: into the United States processed-meat pipeline. How do I square that fact with my ambition to be a meat-eating environmentalist? I can't, is the truth.

To embrace sustainable farming along the lines espoused by the IPCC is to slow desertification and degradation, drop the waste, enhance soil fertility, increase carbon storage in the soil, and create a foodscape that's secure and resilient. It's hard work. So, most people who responded to Roberts' comments just started eating plant-based foods. It's an easy out. *Go vegan*, the choir sang. *Or at least flexitarian*, the less committed replied. Flexitarian — a word I've started dangling in front of my dad, as though it's a shiny new trinket to be admired as explanation for why beef is disappearing from our table after only months as cattle farmers.

The term flexitarian was coined in the 1990s but given fresh life by Marco Springmann, the lead author of an Oxford University paper published in *Nature* in the wake of the IPCC report that called for less beef and more plants in the Western diet if we were to continue to produce enough food for the growing global population within 'planetary boundaries'.[17] Nature has limits, Springmann argues, and in its use of land and water, the modern food and farming system is pushing right up against them.[18]

The flexitarian diet suggests an adult can fulfil their nutritional needs by consuming a mere 14 grams of red meat per day, or 98 grams a week.[19] In some parts of the world, that requires cutting red and white meat consumption by more than half. It means a small steak once a week, rather than a large one four times, and includes small amounts of fish, chicken, and dairy, plus a lot of legumes, fruit, and vegetables. Reduce the amount of beef and dairy in our diet — that's it.

Reduce.

Can it be that simple? No. It soon becomes clear to me that the flexitarian guidelines have been read as an attack on a way of life, an identity, a culture, and, importantly, a large, profitable, powerful industry.

There is consensus among the international science community that carbon emissions must drop, if not cease. Under the Biden Administration, the United States, one of the largest emitters along with China, ambitiously committed in 2020 to cut carbon emissions by at least 50 per cent below 2005 levels by the year 2030. But the immediate focus is on methane. The United States and European Union are leading a global pledge to drop methane emissions by 30 per cent below 2020 levels by 2030.[20] Anthropogenic, human-made methane emissions are an easy target — the oil and gas sector's fracking industry is a massive contributor to the global methane burden via leaky wells, and these can be capped[21] with political will and regulatory pressure.[22] But how to best deal with biogenic methane emissions from belching cattle, rice paddies, and wetlands is more contentious. Currently, the agreed global standard is to calculate methane emissions as a carbon equivalent (GWP100) to roll it into offsetting targets and trading mechanisms.

In Australia, farmers have been moving faster than the government for years. The National Farmers Federation, a membership-based policy and advocacy group, is pushing the sector towards a carbon-zero future, and not waiting on the government to act. In 2021, then prime minister Scott Morrison resisted global calls for Australia to adopt ambitious targets, declaring instead that the realistic goal for the country is a 26 to 28 per cent drop below 2005 levels by 2030, in line with the Paris climate agreement. New Zealand's Paris Agreement pledge is to drop emissions by 30 per cent below 2005 levels by 2030, and to hit zero carbon by 2050. Australia's stumbling block is energy — the country is heavily reliant on fossil fuels to generate electricity and to power transport and manufacturing.[23] It has a heavy carbon-dioxide footprint to erase. Whereas New Zealand is unusual among developed nations, because almost half of the country's gross emissions come from the methane-heavy agriculture sector. It's all the cows.

Under global agreements, biogenic methane is calculated as a carbon equivalent, even though the two gases behave very differently. Methane is a potent, short-lived gas that only exists in the atmosphere for 12 years, meaning that reducing methane emissions can actually cause cooling. By contrast, carbon dioxide lasts around 1,000 years, so the fossil fuels we burn will keep accumulating in the atmosphere.[24] We can make gains now by cutting methane emissions, but that will do nothing for future generations living under the burden of our carbon-dioxide addiction.

Professor David Frame, Director of the New Zealand Climate Change Research Institute and lead author of the IPCC 'Fifth Assessment Report', has long advocated a shift away from emission percentages and the packaging of measurable carbon-dioxide equivalent equations to a conversation about *warming*. A difference made clear in a new equation for measuring emissions: GWP*. But the international community haven't adopted it, and New Zealand's Minister for Climate Change James Shaw told me there is no appetite to reopen negotiations over measurement equations.[25] It took too long for the international community to settle on GWP100. Nonetheless, Frame proposes a two-basket approach to account for the different warming properties of short-lived biogenic methane and long-lasting carbon dioxide.

'The way I see it,' says Professor Frame, 'we should be having separate conversations about those gases which need to get to zero [carbon dioxide] and those gases which do not.'[26] We're not, though. The global conversation is focused on translating all greenhouse gas emissions into carbon dioxide equivalents to achieve a carbon net-zero goal. The problem is the complexity of it all and the co-opting of the messaging in pursuit of profit and politics.

Professor Frame is a pragmatist, a climate scientist, and an imminently qualified physicist who has been the lead author on two Assessment Reports for the IPCC — the documents that keep me up at night. He spent much of his early academic career at Oxford University, and he considers the current conversation about global warming an exercise in neocolonialism. The heavy carbon-emitting European Union, for example, wields enormous trade power over once colonised regions that produce a high level of biogenic methane per capita, like the rice-growing nations of Southeast Asia and cattle-heavy New Zealand. Neocolonial power is still imposed via economic policy and trade, and it filters into climate discussions through the simple act of the European Union providing a disproportionate number of authors of global climate policy.

'Dave from Invercargill', as he describes himself jokingly, has no slavish devotion to the European way of doing things, because he has a better idea for how to tackle the issue of warming in this part of the world. It includes treating gases and their warming potential differently, and acknowledging that, worldwide, cattle are not the only, or even the main, problem to solve. He says plainly that we can't afford another two generations driving vehicles powered by fossil fuels. The planetary reserves will be spent if that happens.

And yet, I'm fixated on cattle burps, Meatless Mondays, and flexitarianism. And I'm not alone.

The global cattle herd numbers around 1.4 billion, according to the FAO. For most people, this is just a number on a page, but for me, the methane percentage point attached to that 1.4 billion represents an animal on my farm and hours and hours of labour.

Am I channelling my lifeblood and savings into rearing animals that are helping degrade the planet? If we cull the global herd, we very quickly lessen the amount of methane released into the atmosphere.

Reduce the herd.

Can it be that simple?

No.

It's nonsense, Hamish tells me. After my uncle introduced me to Bill, I don't stop asking questions, and eventually I find my way to the table of Hamish Bielski. He knows my cousins, who also farm, and he knows a lot about carbon cycles and methane. That's how farming communities work: everyone knows someone who knows something. And once you're accepted, the sharing starts.

He meets me at the door of his farmhouse wearing a heavy woollen jumper, green fleece trousers, socks, and slippers — the farmer uniform — and we settle into conversation at his dining table, still set from the breakfast his three children have eaten.

Bielski came up through the farming ranks as a shepherd and stock manager, and he knows what it is to farm intensively and 'conventionally', but he's focused now on an alternative way of farming. He's trying to decarbonise; to change the way farms work entirely. Aboveground carbon is stored in living trees and shrubs, in coarse woody debris, dead trees, fallen woody litter, and leaves; below ground, it's in the soil as organic carbon. To farm without disrupting all this carbon is the goal. It's farming *with* nature, not against it.

As we start talking, Bielski says he's frustrated with the global community's fixation on slashing methane emissions, because it doesn't incentivise full system change within the farm boundary. Its focus is on offsets — which is, essentially, a process of accounting. Farm or corporate emissions are emitted in a specific location and then carbon credits can be purchased in, for example, Sri Lankan solar projects or permanent pine-tree forests in New Zealand or the United States, and the equation is balanced back to zero — or, at least, that's the goal. The scheme has its detractors, who, like Bielski, point out that offsetting means to avoid action that will entirely change how we do things and wean us off fossil fuels.[27]

For example, Yarra Biodiversity Corridor in south-western Western Australia has planted 29 million native trees across 100,000 hectares to create a carbon sink and habitat for threatened species. These trees give companies like Carnarvon Petroleum, Swarovski, and Scottish craft-beer brewers Brewdog[28] permission to maintain the status quo rather than change the way they mine, trade, or brew. Since 2018, the New Zealand government has funded landowners to plant out 27,640 hectares in exotic and native trees under its One Billion Trees program. It's land locked up in order to create carbon credits that are traded as units under the government's Emissions Trading Scheme. New Zealand's ambition is to reduce methane by 10 per cent below 2017 levels by 2030, and up to 50 per cent below 2017 levels by 2050. And to do it mostly via stock-number reductions, which are a natural consequence of farmland being sold to carbon farmers, who plant entire properties in trees that will work as carbon sinks.

New Zealand is one of the countries that supports the United States and EU's pledge to cut total global emissions by 30 per cent by 2030. Australia refused to commit, describing the ambition as 'extremist'. Then minister for Industry, Energy and Emissions Reduction Angus Taylor said that the only way for Australia to reduce methane emissions, half of which stem from agriculture, is to start culling cattle, and he won't do it. 'What activists in

Australia and elsewhere want is an end to the beef industry,' he wrote.[29] Never mind that a third of the methane burden is imposed by the oil and gas industry.

What's wrong with culling a few cows! I yell at the minister. *If the result is cooling?* But that's my default green city leftie view. A position built on outrage.

I realised not long after moving to the farm that *I* was the problem I was trying to solve. In my mind I'd already written the headline — 'Our addiction to big beef has ruined the planet!' — and made the protest sign: 'It's not food, it's violence!' I'd imagined donning a white boiler suit and, with my iPhone aloft, storming a feedlot, unlocking gates and sending 600-kilogram cattle rampaging down the dirt corridor.

I accused dairy farmers of poisoning our waterways and dumping effluent into the streams. I built myself up with accusations. I told myself I was caught between the culprits and saviours. And I waited for someone to tell me what to do, who was responsible, and how I could help stem the wave of environmental change. I was waiting for the most convincing story.

But there is no one person to blame. No one person with the solution. No one solution to our collective problem of global warming. And so I listen as Bielski tells me that cows are not the problem — it's how we're farming them.

I try to keep up as he starts talking about plant biology and biogenic methane and carbon cycles as explanation for why he thinks calls to drop stock numbers are more about politicking than curbing global warming. He explained the carbon cycle to me like this: plants need carbon dioxide to grow; they draw that from the atmosphere through photosynthesis; cattle eat the plants then belch methane as part of digestion. Methane lingers in the atmosphere before converting back into carbon dioxide for plants to draw back down. So, he reasons, if a farm has the same number of cattle now as it did in, say, 1980, the same hypothetical four tonnes of methane would exist in the atmosphere because it's constantly recycled through the carbon cycle.

Unlike fossil fuels, which add layer upon layer of carbon dioxide to the atmosphere, livestock do not compound the methane burden *so long as the carbon cycle remains unbroken* and the farming doesn't intensify further. Soil undisturbed, plants in place is the solution. Don't cull the cows. Change the farm. Some see this ambition as retrograde, dismissing it as 'old-fashioned' minimal-input farming, much like what Bill does on his steep hill country farm. But it's not anachronistic. It's forward-looking in that Bielski, and others like him, know the farm is part of an ecological system in flux; the future is not today. The challenge is to get comfortable working with the flow rather than trying to stop or control the movement, as farmers tend to do when they follow a system that assumes or demands they produce the same crops, milk, beef, or lamb in the same time frame, year on year, regardless of the conditions.

You don't need to look too far to find examples of cattle ranching and farming that make a claim for carbon neutrality and environmentally friendly practices without changing the system. In Australia, the North Australian Pastoral Company[30] — which farms 200,000-odd head of cattle across 6 million hectares — sells carbon-neutral beef and is able to do so because it offsets its emissions by buying carbon units in renewable energy initiatives like the Danjinghe Wind Farm Project and the Rice Husk Thermal Energy Generation Project in China.[31]

To me, offsetting looks at first glance like branding shenanigans that avoids on-farm improvements. It keeps the old system intact, and it makes city people distrust farmers because they're buying their way out of the cow burp quagmire, and charging consumers more for the result, which is carbon-zero beef. I'm no fan of this trickery. It feels like cheating.

Hamish leans back in his chair, locking his fingers behind his head. 'I may be wrong,' he says. 'But I don't think I am.' What we all need to do, Hamish says, is drop our entire energy use. We need to decarbonise.

We need to shift away from the way most farm now, with a focus on heavy cropping cycles, buying seeds, fertiliser, fuel, more seeds, fertiliser, and fuel, all to grow winter feed crops and new grass in succession. Yet for the entire industry to change, Bielski reckons, it'll take 20 years.

Twenty years. *That's too long*, I think to myself.

'My problem is I go too fast,' he says. Bielski is proving that one can farm with minimum inputs in a manner that protects the ecosystem, but his farming methods are pushing at a collective bruise caused by the public labelling of farmers as polluters, and he's sensed a cooling in his community.

'No one wants to talk to you because "you're extreme", "you're an ideological person",' he says.

'*Are* you ideological?' I ask.

'I don't think so,' he says, before elaborating that his motivation is actually quite modest. 'I just want to disrupt the status quo.'

I leave Hamish's warm home, thinking him less an ideologue than a sage. Calm descends as I mull over his words. If I change everything we do on the farm, will I, too, be seen as an ideologue — worse yet, an activist? Is that a bad thing?

I fear the cooling effect from community. We — Pat and I — are already alone, removed from the life we built for ourselves in Sydney. To be ostracised even further seems … dangerous. But there is another cooling effect that's bigger than us. Achieving a significant drop in methane emissions can have a cooling effect on the planet. We have a chance to stay below the 1.5 degree increase that the scientific and political community is focused on. It's a Band-Aid to slow global warming. *Drop the methane now. Decarbonise next.*

Rebuild community later.

Yes! I scream out the car window to no one. *That's doable.* But my renewed energy doesn't last long.

I wish Pat were with me but he's at home, moving cattle, doing the work. I call him. 'Did you learn anything?' he asks flatly. 'I did, I did!' So much.

Then he shares his news. His 11-year-old niece has been marching in the streets, surrounded by signs showing the planet broken and bleeding, and shouting 'Stop Denying the Earth Is Dying', 'Killing Animals is Killing the Earth', 'Flight Climate Change with Diet Change'. I wonder if she's embarrassed to tell her friends that her aunt and uncle have left their inner-city Sydney life to become cattle farmers: environmental enemy number one, the opposite of vegan.

2.

Silver linings, silver bullets

When I was small, I would walk barefoot along rows of silverbeet, leeks, cauliflower, and broccoli, catching hovering white butterflies and tiptoeing around the sheep droppings scattered by my mum at the base of the vegetable plants to help them grow. I was taught to cook at a young age. It was simple cooking — fresh whole food was steamed, baked, stewed, or roasted, seasoned with salt and pepper, and presented as a meal. The flavour wasn't an additive — it was in the vegetables that, minutes before cooking, had been in the soil.

I was never reprimanded for my raids on the vegetable garden. I'd rip stalks of silverbeet from the plant to combine with lemon juice from the fruit off the tree. This brew would be sold at my imaginary food stall alongside a chocolate cake I'd whip up following a recipe of my own design that featured more salt than sugar.

My dad was the reluctant taste-tester, my mum the tutor. I fell on the stovetop around this time and burned a neat mark into my stomach; forever branded a cook. This was the 1980s, and the age of 'meat and three veg' cuisine. And I've eaten meat since, until the weeks I nursed calves back to life or lost them to dehydration. When I realised I was essentially factory-farming. Then my meals changed.

The death on our farm caused me to lose my appetite. Death takes a toll. I've learned that in many ways over the past year. And its association with food doesn't make the phenomenon any less confronting if you dwell on it.

Death trundles down my road each day during the spring calving season, which can start as early as July, in the depths of the Southern Hemisphere winter. It's represented by small stock trucks that arrive at dairy farms across New Zealand, collecting bull calves destined for slaughter.

'Consumers do not think about the fact that milk and dairy come from a cow, much less come from a female cow, much less come from a female cow that had to have been pregnant and given birth in order to be producing milk,' says Sophie Egan, food writer and program director at The Culinary Institute of America. She had been thinking about this fact when we started corresponding because she's a breastfeeding mother. It dawned on her that advertising for milk products usually obscures the udder, the teats. People don't want to think about where their milk comes from, let alone how it's produced.

It's shocking to some that a dairy cow is often impregnated through a process of artificial insemination. Two hundred and eighty days or so later, the cow will give birth to a calf. If it's a bull calf, odds are it will be slaughtered. If it's a heifer calf, it might be reared by workers on the dairy farm and then put into the milking herd when it's old enough to have a calf of its own. This is the process required to keep cows producing milk, mostly for human consumption.

It wasn't that long ago that normalcy was shooting a newborn bull calf in the head with a rifle. Now, governments across the countries connected by the legacy of the colonial food regime have introduced regulatory frameworks designed to limit a calf's suffering in its short life. Just under two million calves are killed each year before they are a week old in New Zealand.[1] In Australia it's about 400,000.[2] And the process is clear: calves must be well cared for in their first days of life. Shelter, warmth, twice-daily milk feeds are all compulsory.[3] Animals must be healthy when they are loaded onto the truck via a purpose-built ramp.

In the United Kingdom the bobby trade has come under fierce scrutiny. For years calves were shipped to the European continent in a live-veal export trade. Animal rights activists lobbied for this practice to stop, and so in 2018 not a single calf was exported from England.[4] Instead, some 95,000 bull calves were shot. Many more were reared for beef, as I do here in New Zealand.

The campaign to value these dairy animals is supported by chefs like Jamie Oliver, who exuberantly encourages Britons to buy sustainable rose veal because the 'higher-welfare product tastes great' and that's worth celebrating.[5] Rose veal is the name given to beef cuts from an animal that's no longer suckling milk from the cow but is still young, under 12 months old. But the thought of calf meat lingers and is repellent to some consumers.

What's increasingly more repellent to me is not the age of the animal, but its very existence. When I started farming, I thought buying bull calves from the dairy industry was a positive thing, because so many farmers considered these baby cattle a waste product. *Make a business from unwanted bobby calves*, I thought. *Prove the farmers who say they're valueless wrong.* If all we're doing is farming cattle for mincemeat, it makes no difference if the animal is male or female, a bull or a cow. But as each day passes, I'm more convinced this decision was misguided. And now a sobering question is settling: should I leave those calves to their fate? A fate that means they leave the dairy farm at four days old on a truck destined for the abattoir. And not a bit of the animal is 'wasted'. The tiny animals are killed for milk veal meat destined for the niche European market. Their hide is turned into the soft leather used for expensive handbags and sports car interiors. The carcass becomes a blood-and-bone mix that helps vegetables grow, and in the blood is newborn calf serum, an ingredient used for medical cell research — the development of vaccines, in particular.

Calves have a use off the land. And isn't that what we should be focused on? The health of the land, the fate of the planet?

In order for us to stay within the planetary boundaries, we're told, we must all eat less meat and dairy. Better yet, none at all. That will result in fewer cattle to burp methane and excrete nitrous oxide via urine and dung, hypothetically. I've watched the life fade from an animal's eyes, turning foggy in an instant, where once I could see my reflection in the glassiness. The image stays with me. I understand the emotion that spurs the cry that we must stop cattle farming. It can be a heartless business.

My niece marched in the streets of Sydney for the School Strike for Climate, following in the footsteps of environmental activist Greta Thunberg. Now her friends are switching to vegetarianism; some hope to be vegan. Veganism is in the news, and the movement once wholly focused on animal rights is now spotlighted as a solution to climate change. My niece and her friends are desperate to take action from the playground, and what's in their lunch box suddenly matters.

My heart bruises a little when I read the signs and listen to her excitedly tell me about her group's experiments with dietary change. Not because I don't share her concern — I can't stop weighing the impacts of intensive animal farming against the potential gain if it doesn't exist. Yet the moral certainty of the position 'go vegan for the sake of the planet' shades so many of the big questions an 11-year-old must face down in the years ahead of her. Questions like: what role food will play in her life? Will she care about provenance, preparation, nutrients, taste, texture? Will food bring her pleasure, or will it become just fuel? And if it's just fuel, what form will it take?

That certainty also obscures many of the questions we must all face: how will we feed a growing population on land that is degrading without degrading it further, in conditions prone to flux, in rising temperatures, all the while reliant on a food system that can be pushed to breaking point in times of crisis? Is it as simple as — cull the cows, plant more crops? Stop eating meat, go vegan?

I know my niece and her friends' collective anxiety can't be diminished. It's so acute the mental health community are investigating. In 2018, researchers from the Australian National University found that more than 85 per cent of surveyed 12- and 13-year-olds believe they need to worry about climate change.[6] That it isn't something the adults will sort out.

To be vegan is to care about the state of the planet at a time when climate action is the rallying cry of young people the world over who are racing towards political awareness.

Vegan vs farmer. It's a polarity exacerbated by our leaders. The former Australian prime minister Scott Morrison saw no harm in labelling vegan activists 'green-collar criminals', while livestock farmers are attacked by animal rights groups with accusations that they're dullards, murderers, and environmental terrorists.

One protestor in Melbourne, 13-year-old William Currie, told Plant Based News that 'going vegan is the only way to stop or at least slow down the process of climate change'.[7] He went on to explain that he will fight for what he believes in, no matter what the leaders decide. 'All rights came to be from people rebelling and striking for what they believe is right and just,' he said.

Perhaps William is right. We must stand up for what is right and just. But if the goal is individual action for collective wellbeing, who decides the right course of action in the climate battle if food is the focus? Me? William? The prime minister?

'Do you think about our cattle after they leave on the truck?' Pat asks one day. After months of hard blue sky and no rain, we've just sent 150 100-kilogram bull calves off the farm to a larger property south of the lake, near where we live.

'Yes,' I reply. But then I stop myself. I can't think of it too much. 'Do you?'

'Not much,' he says. *That's the difference between you and me*, I hear him think. *You're trying to solve everyone else's problems; I'm just trying to solve ours.*

The rain comes fat and heavy later that day. As the afternoon wanes, black, boisterous clouds rumble over the steep mountain that spits hot geothermal steam from a scarred face at the back of the farm. The front creeps low along the waterway, settling into the valley, and then opens.

It's like the rain we knew in Sydney — ferocious storms that would barrel in from the red centre, bringing golf ball–sized hailstones and flooding roads, only to vanish over the ocean horizon 30 minutes later. We're not used to it here.

Water is likely to be how we'll all experience the impacts of global warming. Through floods and drought, mostly. Concentrations of carbon dioxide, methane, and nitrous oxide in the atmosphere have increased, and the consequences of the current 1.1 degrees Celsius surface-temperature bump is becoming easier for us to recognise. It presents as heatwaves, storms that are more violent, droughts that are more prolonged, and floods[8] caused by river systems that reclaim centuries-old pathways through prime farmland. The tepid, late-autumn drizzle of a decade ago has given way to dry blue days and monsoon rains. New Zealand now has Australia-like storms so heavy that the water carves new rivers down farm tracks, and floods towns where stormwater drains have been built for the polite weather of yesteryear.

As the rain blows off, a golden evening peeks through the bruised clouds, lighting the paddocks and water in a warm, clean hue.

I am pregnant again.

And so fearful.

Not of losing her, but of everything else. The rain. The flooding. The fate of our cattle at the farm where they're heading. And the chasm that's widening each day between us — Pat and me — and between our imagined farm life and the reality; between the people who grow food and those who consume it.

Pat pulls on a rain jacket and hand-me-down gumboots with cracked soles and walks out the back door, off to check on the small number of calves we have paddocked behind the shed. I stay home and read. 'Failure to implement animal to plant-protein shifts,' writes American researcher Dr Helen Harwatt, 'increases the risk of exceeding temperature goals.'[9] 'Evidence indicates that food production,' writes the influential researcher Professor Walter Willett and 36 co-authors in *The Lancet Commission*, 'is among the largest causes of global environmental change.'[10]

Is that what we're doing on our farm? Causing environmental change? For the worse?

I have started cooking my favourite berbere beef recipe with canned lentils, imported from Australia. Vegetable stock and lentils replace beef stock and meat. The cooking time is cut by a third. I sprinkle coriander atop the flavoursome dish and place it in the middle of the table with tangy flatbreads made by combining wholemeal flour, buckwheat flour, and salt with yoghurt until it forms a stiff dough. Rolled flat and cooked on a hot, dry pan, the small bread rounds become a utensil for mopping up the sauce.

When Pat returns, we sit opposite each other at the table and rip the bread to form scoops to capture mouthfuls of stew. It's deep brown, hearty, and rushes warmly to the searing parts of my stomach, hot from the ever-present burn of anxiety. My cooking without animal products has become an exercise in mimicking the umami-rich flavour of animal products. It's a nostalgic trigger that makes food more than fuel for me. In the tingle of a spicy gravy, a feeling of joy emerges that, just for a moment, squashes the sadness.

'This is really good; I like it better without the meat,' says Pat. 'You've always wanted to eat more meat than me.'

He's right. I have a deep want for something that red meat satisfies. I've always thought I needed it more – for energy, warmth, comfort. And something else indefinable. But that desire is dissipating.

There is a common origin story among the vegan animal-rights activists I've met recently: 'If people saw what I saw, they'd change. If people knew what I knew, they'd do something.' Do something, like protest, lobby, shop differently.

Peter Singer, the Australian academic and author of the seminal animal rights book *Animal Liberation*, tells me over the phone one day that he has vivid memories of protest actions that pinned a bullseye on the backs of the political elite. Trained at Oxford University in the 1970s, Singer continued his animal rights activism in Australia in the years after, and recalls picketing, in 1992, a Hunter Valley piggery part-owned by former Australian Labor prime minister Paul Keating. The protesters chained themselves to stalls and fences and called the media, determined to bring attention to the crowded housing conditions of the breeding sows and litters. It was a playbook move by the Animal Liberation group, who continue to this day to elicit media coverage of their protests.

Australian and UK farmers and feedlot owners fear these actions. One New South Wales feedlot owner, who feeds 6000 cattle housed in a long row of square open-air pens, explained that she has trained her staff to manage (not inflame) the situation if protestors arrive at her property. Her priority, she says, is keeping people safe around the constantly moving trailer trucks, heavy machinery, and even heavier animals. 'If a hundred people turn up to your property yelling and chanting and calling you names, you essentially can't move them off,' she tells me. 'So you become a tour guide and you try to control their movement through your business.'

I imagine how the protest unfolds: the activists bellow at their audience watching the action through a live Facebook or Instagram feed: 'Just look at what they're doing; going vegan is the only way!' The farm workers scream back: 'We're doing nothing wrong; this is what the world needs to stay fed!' And real, identifiable, large-scale and lasting environmental change remains elusive.

Why?

In part because protest images are evocative. It's hard to look away from the sight of tens of thousands of animals penned up. It's hard to ignore a good story in which the villain and hero are so clearly rendered. But that's the moral battle; environmental change is the fuzzy backdrop. It's hard to define, not always pretty, and often stems from boardroom decisions about profit.

The influential food theorist Professor Harriet Friedmann writes that big food companies have long dictated what we eat. Cues may be taken from protest movements like Peter Singer's piggery protest and Greta Thunberg's global School Strike for Climate, but change will be managed under the umbrella of corporate environmentalism[11] to ensure profit margins remain intact. And for the most part, people are happy to go along. 'Shop to stop climate change' is a motto easy to embrace.

At the core of consumer-focused change is the green capitalist's checklist.[12] Tap into the consumption habits of wealthy consumers; develop targeted supermarket retail strategies; focus on environmental rhetoric; develop complex audit processes whereby standards are established to ensure a farm or manufacturer operates to a certain level; enforce inspection and traceability; and, finally, champion emblematic new products. The point of an iconic product offering and marketing campaign is to fill a void, especially an emotional one, with something.[13] The current 'void' is anxiety over global warming; the solution is anything labelled vegan.

Big food followed this path with free-range eggs off the back of consumer concern over animal welfare. The same strategy is being fervently pursued for the vegan food market. Consider the small shifts over the past few years: vegan foodstuffs are marketed to the wealthy consumer who has been branded by industry as the 'conscious consumer'; supermarkets now display plant-based meat and milk replacements next to the animal-derived equivalents; 'veganism will help the environment' is a catchcry often heard in mainstream media; new global standards for assessing environmental impacts are being added to product labelling; and the emblematic products leading the charge are plant-based burgers. The checklist is complete.

In the rush to 'do good' — or, at least, to be *seen* to do good — consumers are being sold an inflated fix by corporations pursuing a green agenda for more profit. But that's nothing new.

In 2018, a trio of Australian academics looked at the motivation for big supermarkets — namely Coles and Woolworths, Australia's retailing giants — to take the lead in pushing for better animal welfare standards in large-scale farming in the absence of government regulation.[14] In the wake of a concerted effort by animal welfare groups to expose confronting practices within the poultry and pork industries, specifically the use of battery cages and sow stalls, Coles and Woolworths adopted a voluntary labelling system designed to push large-scale producers into free-range practices, because that's what consumers were demanding.

Woolworths started to phase out its own-brand cage eggs in 2009; Coles followed suit in 2010. By 2015, the majority of carton eggs in Coles and Woolworths were free-range[15] — a term used to describe chicken farms that give the birds daylight access to an outdoor area. This does not mean an idyllic vision of hens scratching through orchards or grassy pasture, though. It's a calculation: eight hours outside, one bird per square metre.[16] And the retail giants proclaimed that they had changed the farming landscape for the better, at no added cost to consumers.[17]

This idea of consumer-driven change or 'vote with your fork' has currency, but whether it results in the level of change the consumer expects or believes is questionable. The supermarkets' adoption of ethical branding like 'free-range' and 'RSPCA-approved', at a price point that keeps the products on the weekly grocery list, gives the impression that animal welfare and low prices are possible at the same time.[18] And yet, voluntary higher-welfare labelling practices don't result in much improvement for most farmed animals. In the case of eggs, the majority of those produced in Australia don't end up on the free-range supermarket shelf — they're battery eggs used in food trade, catering, restaurants, and processed foods.

As for pork, around 70 per cent of the sow herd is sow-stall free — i.e., no longer housed inside a two-metre by 60-centimetre metal bar crate during pregnancy — but only 5 per cent is in free-range systems.[19] It's 'welfare washing', according to Australian food system researchers Christine Parker, Rachel Carey, and Gyorgy Scrinis.[20] If there is choice — a range of products at different price points with a scale of animal welfare labelling — then there is the illusion that voting with your shopping trolley works and change is possible.

We can see the same consumer pattern happening again, but this time it's vegan washing.

Australia may have once been a nation of red-meat consumers, but, as in New Zealand, it's a habit on the decline. Australia is now one of the fastest-growing vegan countries in the world — if by vegan you mean those who subscribe to the diet sometimes, casually, and often with their wallet and Google.[21] In 2019 and again in 2020, Chef's Pencil claimed in their 'Vegan Food and Living' report that by analysing Google search trends, they could discern that Australia and United Kingdom were the top two fastest-growing countries for veganism, with New Zealand at number five, because lots of people googled terms like 'vegan recipes' and 'vegan restaurants'.

Supermarket chains backed this claim, especially in the United Kingdom, where food market analysts Mintel report that 65 per cent of Britons tried a meat-substitute product in 2019.[22] Nearly a quarter of all new food products launched that same year in the United Kingdom were labelled vegan. But for every ethical gain, such as a win in animal welfare, another part of the supply chain may suffer, as I was to discover.

I'm given steel-toed cream gumboots to wear as I enter the Fisher Meats Factory on the eastern outskirts of New Zealand's southern city of Dunedin. Size 39 and slightly too big. Grant Howie, a former meat-processing executive turned plant-based product maker, hands me a white lab coat and a hairnet, and tells me about the safety procedures for the factory. I'm to stay close, watch out for forklifts, stay away from the machines and from the men and women with sharp knives. It's a 'look, don't touch' tour.

He's let me inside Fisher Meats because I want to learn about making plant-based sausages. My farm life has put me off red meat, and like so many others around the world, I'm starting to appreciate the no-harm claims of the vegan diet. But I'm wary of meat replacements. The ingredient list of these products reads like a science experiment: pea protein, soy protein, wheat protein, coconut oil, oats, barley, hemp, salt, vinegar, natural flavour, ascorbic acid, lactic acid, gluten flour, vitamin B12.

In the deep south of New Zealand, Fisher Meats is respected as a legacy brand. It's a company that has employed Dunedin workers to break down beef and sheep carcasses for more than a hundred years. But in 2018, alongside the French lamb racks and beef cheeks packed for sous vide, they started making plant-based food — like a chorizo sausage that's not really sausage.

If I start advocating an end to the meat and dairy industries in the name of the environment, I'm threatening the old protein trade — animal products — but not the new. Not the sausages Grant is producing. And he's not alone. Beef and dairy companies the world over are switching allegiances, away from their farm suppliers to food start-ups that make protein, be it plant-based meats, yeast-fermented cheese, or cultured steak and chicken nuggets — the lab-grown meat of our future. And they're increasing their profit margins, because plant-based protein products command a premium price, selling for a couple of dollars more per kilogram than rump steak.[23]

American food writer Sophie Egan believes that the West tends to value the new and the novel, especially when it comes to food. In Sydney, Pat and I rarely spent time during the week cooking, eating meals slowly at a table, or eating with others. Food was taken on the go or squeezed quickly into the schedule between 7.30 and 8pm. Following food trends via takeaway or pre-packaged and ready-to-make meals is intimately connected to a narrative that productivity, and work, is good and essential.[24] 'It means we're willing to try all kinds of different cuisines and different food,' says Egan, elaborating that it also means that we are overvaluing food products and stunt foods that come from labs. In 2019, Barclay Bank investment analysts suggested that the alternative-meat market was on track to become a US$140 billion industry within the decade, swallowing 10 per cent of the global US$1.4 trillion meat trade.[25]

Dressed in white boiler suits, Grant and I descend concrete stairs painted yellow and step into a room of stainless-steel equipment, where a faint hiss is audible and a chalky odour hangs in the air. In the corner is the sausage maker, scrubbed clean and shining. Not so long ago this room would have pumped out hundreds of slender beef snags made from something approximating pink slime. Now the slime has been replaced by dry ingredients that are hydrated, flavoured, and made rich with oils.

Making plant-based sausages is like baking. There's hemp powder from Canada, coconut oils from the Pacific Islands, soy protein, pea protein from other parts of the world, and flavour essences that come from corn and barley. 'So, you're trying to approximate the flavour of beef?' I ask Grant.

'Yes, correct. Both in texture and flavour profiles.'

The Craft Meat Company's plant-based products may have vegetable illustrations on the packaging, but the content of the food stuff is mostly plant powders fortified with vitamins and minerals. My toes are numb inside the cream gumboots and my nose has started to run as we leave the sausage area and descend deeper into the factory, to where the red meat is processed. It's two degrees colder inside the plant than outside, and it's only six degrees outside.

The plant's workers are also wearing white boiler suits and layers of warm clothing beneath, making them into a small army of marshmallows. They stand around the processing tables, joking with each other, flicking their knives to Womack and Womack, removing flesh at a rapid pace. I ask Grant what he means when he says his company is trying to be more sustainable. Part of it is about lowering their carbon footprint. 'When you compare beef's greenhouse gases to plant-based production our plant-based produce has a much lower footprint,' says Grant, despite the fact that most of the ingredients are imported from Canada and Europe,[26] large producers of plant-based isolates.

In a meta-analysis of Life Cycle Assessment (LCA) studies conducted in Sweden in an effort to clarify questions on the environmental impact of plant-based foods, the researchers concluded that plant staples such as dry or canned beans and peas, even after import into Sweden from legume-growing countries like Canada and Australia, show a climate impact below 1 kilogram CO_2 equivalent (CO2e) per kilogram of product.[27] The impact goes up for plant-based protein products like Grant's sausages. Quorn made from wheat-based sugar, for example, amounts to 2–3 kilograms CO2e per kilogram of product. In comparison, the global average for beef sits somewhere between 27 and 36, depending on the study.[28]

Yet as the Swedish researchers point out, location makes a profound difference. A smaller CO2e number doesn't account for water use, water contamination, biodiversity loss — it's not as simple as saying plant-based foods are always better than beef if you don't understand that some plant-based foods made from isolates have a heavy energy footprint at the point of processing, and some plants, like fava beans grown in arid Egypt and exported to Sweden, have such a heavy blue-water burden as to render them a far more impactful food than beef production in the country importing them.[29] And yet, the fava bean crop is grown in abundance, under irrigation, because it's an essential staple in the Egyptian diet as well as a valuable export income for growers.[30]

I wonder how Pat's new plan would stack up against these numbers. 'There's a feed supplier making calf meal just north of us,' he tells me excitedly. They sell it by the tonne in large sacks. 'But it's got biscuit crumb in it.'

What? Biscuit?

I learn later that it's a waste product from a biscuit factory and when crumbled into a mix of grains it gives a sugar-rich energy boost to the calves in the depths of the cold winter.

'Should we be feeding our calves biscuit?' I ask.

'I don't know,' he says. 'But it's cheaper, we get rid of all the individual feed bags, I don't have to pay for freight. It makes sense.' He's right, it makes sense for us, now. The fact that we can doesn't mean we should — but we do, because food is not just a CO2e number.

Grant says sustainability must be about three things combined: the environment, the labour force, and finances. He's entered the plant-based product game for the New Zealand and Australian market because he knows his customers are 'conscious consumers', which is to say, they shop with the environment in mind. And to ward off imported products, he's expanded his protein game because the plant-based trend is rocketing up. If he sells more products across the protein spectrum, he can keep his local staff employed. Every effort to end beef and dairy, an industry that employs directly or indirectly around 115,000 people in New Zealand and near 200,000 in Australia,[31] comes with a job put at risk unless something like plant-based manufacturing replaces it.

As we talk, Grant continues to use the word protein rather than meat, beef, cheese, or milk. It is a prized word in the food industry. Humans need protein to survive: it's the macronutrient that helps us to grow, and without it, malnutrition looms.[32] But the West has become obsessed and anxious about its presence in our food. Most supermarkets will display an array of protein-enriched products from sweet fruit-and-nut protein balls to smoothie flavouring. Beef is a complex protein food source rich in minerals including iron, zinc and vitamin B12. When the animal is fed on grass, the beef is also found to be higher in vitamin A and E content, with a healthier omega 3 to omega 6 ratio than grain-fed meat.[33] But these minerals can all be added to plant-based foods.

Technically, I'm a protein producer, but beef is the least fashionable food item in the protein family in the current climate. I distrust the word 'protein' because it seems to strip all the cultural and emotional value from the food that comes from animals, rendering it merely a commodity to be traded and added to processed stuff.

A marketing executive named Rick Walker told me plainly that food companies have a vested interest in moving people towards new plant-based proteins because there are more profit margins in it. Rick works for the Japanese-owned New Zealand meat-processing company ANZCO, so he has an interest in spinning a pro-meat line, but the investment strategy of companies like America's Tyson Foods and New Zealand's Fonterra supports his opinion.

Tyson Foods was one of the first investors in the early plant-based burger start-up Beyond Meat, but sold its 6.5 per cent stake at the beginning of 2019, claiming it was intent on starting its own alternative protein lines. This is a company that sells, as of 2021, US$17 billion worth of beef each year and is responsible for producing one of every five pounds of meat (all kinds) consumed in the United States.[34]

In 2019, New Zealand's dairy giant Fonterra, the largest dairy exporter in the world, took a minority stake in Motif FoodWorks, a food design company that makes ingredients such as protein from genetically engineered yeast strains for manufacturers to use in plant-based meats and alternative dairy products. Fonterra was joined by Breakthrough Energy Ventures, a fund investing on behalf of Amazon founder Jeff Bezos and Microsoft founder Bill Gates, among others. They're banking that the Boston-based start-up will succeed in bringing bioengineered food ingredients to market at a scale that will rival conventional dairy ingredients.

For Fonterra, investments like this are about future-proofing. At heart, the company may be about dairy, having formed from the amalgamation of farmer-owned co-ops, but its main earner is milk powder that's used as an ingredient in food manufacturing. If yeast has the potential to truly undercut cows, Fonterra's investment in what it calls complementary nutrition ensures that the company can keep pace with consumer shifts and continue to pay dividends to its farmer shareholders long after the marginal farms are liquidated because it's become too costly to farm cows. In November 2020, the Reserve Bank of New Zealand expressed concern that some 11 per cent of dairy farm bank loans were either stressed or in default. They noted that these farms were reliant on high milk prices just to remain operational. When the milk price crashes, as it always does because the commodity market runs in cycles, some farms will go under. And perhaps they should.

Matt Gibson, who co-founded the San Francisco-based alternative-dairy company New Culture, believes his company, which makes yeast-fermented casein to rival dairy products, is at the vanguard of one the most 'important food movements of our time.'[35] Kraft Heinz's venture capital arm Evolv Ventures dropped US$3.5 million into the seed fund of New Culture in September 2019, and after raising another $25 million in 2021, New Culture is preparing to launch co-branded pizzas featuring its fermented mozzarella in the US market in 2022.

For the big food companies, the movement Gibson speaks of is an upward trajectory of the bottom line; for consumers, it's a bigger bill at the end of the weekly shop. For the environment, the benefit is yet unclear. The CEO of Fonterra Asia Pacific, Judith Sales, says the fermentation process required to turn yeast into dairy may not be less environmentally impactful than dairy farming.[36] The yeast requires a food source, she says. Like sugar, which must be grown somewhere. 'It needs land to produce it and there's waste to be disposed of at the end.'[37] Will these alternative protein products shift the dial on global warming? Are they even intended to?

You may read this and think my bias is clear: that I'm using LCA numbers and investment strategies to create an argument to continue farming cattle. But I have no blind fealty to this life. What I want most are answers to the question that keeps lingering: does beef and dairy need to die for the planet to live? In the very act of asking this question, I've been branded both an advocate for the meat industry and a vegan activist. It's easy to judge and label. But those who judge don't wake each morning and try with every gesture, decision, and act to tread lighter than the day before. To create a life that isn't degrading the environment. It's a messy process that's hard to categorise.

Am I a red meat advocate? Yes, in part.

Am I a vegan activist? Yes, in part.

Am I a meat-eating environmentalist? Yes.

It was late spring when I travelled to the southern tip of New Zealand to meet a young dairy farming couple, but it felt like winter. First came the light snowfall as I drove south, followed by driving rain and then drizzle. In paddock after paddock the soil was bare, some gorged by washout where the rain had pooled into a stream and rushed straight towards the road, carrying topsoil to nearby waterways. In parts of Australia, soil is being eroded by water ten times faster than it's being created.[38] New Zealand likewise loses nearly 200 million tonnes of soil a year,[39] and it clogs streams and washes out to sea, carrying with it the remnants of the nutrient burden applied to heavily fertilised crops and grass. The health of the water is as urgent an environmental challenge as greenhouse gas emissions. Almost all of New Zealand's rivers contain pollutants that are, in small and dramatic ways, changing the ecosystem. Two-thirds of the 1,500 river sites monitored by

Land Air Water Aotearoa show high traces of *E. coli*. Sediment, nutrients, chemicals, and waste are washing into our freshwater from cities, industry, and farms alike.[40] And much of this contamination is entirely preventable. The mitigation advice is bafflingly simple: erosion drops when you keep the soil covered with living plants.[41] Why is this advice so hard to follow?

As I pull up to their farmhouse, the young couple are eating a lunch of scrambled eggs on sourdough, the eggs sourced from the paddock where their chickens roam free-range. They tell me, over a cup of tea, that there have been moments when they've been embarrassed to admit they are dairy farmers. There is some truth to the label 'Dirty Dairying' — a phrase used to spotlight the freshwater and ecological damage caused by the intensification of the dairy industry in New Zealand.

They have friends who have embraced a vegan diet for environmental reasons. Such acts force them to consider why eating meat and dairy is repellent to some. Unlike the farmers shouting across the chasm in a misguided effort to protect the status quo, this young couple has used these conversations as a guide to change.

They don't use synthetic fertiliser, pesticides, or herbicides; they've dropped their herd number from more than 800 to 650 cows; they've embarked on an ambitious tree-planting regime to create biodiversity corridors through the farm; and they don't crop the land. They farm with an ethos that includes the directives *keep the soil covered*, and *plants remain in place*.

Their dairy farm looks different to the neighbouring properties, most notably because of the half-a-dozen long, tall rows of compost, which contain the waste from the calf sheds and calving pad. Once broken down, the compost will be spread on the poorly performing paddocks that run adjacent to the main dairy farm, which is full of robust plants; the grasses and herbage are deep green and voluminous. The other block was once used to grow crops. Every season, year after year, the land would be cultivated, fertilised, cropped, and cultivated again.

In this part of New Zealand, with its runoff-ready soil and heavy rainfall, farmers have started growing oats in larger volumes to supply the flourishing oat-milk market; yet more are growing vast amounts of crops to feed cattle. One crop is seen as planet saving, the other degrading, yet both follow the same pattern: apply herbicide, plough the land, plant seeds, fertilise, spray for pests, fertilise again, harvest. And repeat.

Oat milk made from New Zealand oats and processed in Sweden has a footprint of less than 1 kilogram CO_2e per litre of milk;[42] the same volume for dairy milk from New Zealand, currently considered to have the lowest footprint of all dairy-producing countries, is just above 1 kilogram CO_2e per litre.[43] The consumer switch from dairy to oat milk may result in an overall drop in greenhouse gas emissions, yet the difference here in New Zealand is minuscule. Meanwhile, on the land, a different environmental footprint will emerge if we don't talk about how the crops are grown, not just whether they're being grown. Keep the soil covered, is the advice — the very opposite of conventional cropping. Substituting plant-based products made from plant crops isn't a silver bullet if the farming of the grains and legumes required to make it relies on a cycle of intensive mono-cropping, more fertilisers, more pesticides, more sediment run-off, and more dust storms.

I first met these young dairy farmers at a farming workshop about soil health. I'd prodded Pat to join the group in an attempt to prove to him not all farmers were like my father: conservative, stuck, critical. The sunken hollows beneath Pat's eyes hold so much resentment, and I can see anger settling. They know this look and consider it a sign of the collective ego reluctant to change. And it can quickly turn to anger.

I know this.

What if we stop farming animals, stop eating meat, plant crops instead of processing beef? Will the joy flood back? Or would we be joining an angrier tribe, shouting back across the chasm towards my dad?

When I was with Grant Howie, freezing my toes off in his meat and plant-product factory, I asked him if he thought consumers could drive environmental change through their purchasing decisions. His answer was 'absolutely'. If his 'conscious consumers' continue to want a plant-based sausage that has a smaller carbon footprint than a beef cheek, the company will continue to make them, and in bigger volumes. 'That's what businesses do. If there's no consumer demand, then it's not sustainable financially.' Months after I visit Grant at the factory, I learn Fishers Meats is no more: he's closed the meat-processing side of the business and sold his shares in The Craft Meat Company to Sustainable Foods, a Kāpiti-based company with plans to grow the plant-protein product line. He made redundant all but three of his staff in the process.

Our food system is a complex web of levers pulled by consumer trends, policy, and free market ebbs and flows, and so it's also a playing field for politics. Yet many families don't have room in the budget to buy these more expensive plant-based, carbon-zero products, regardless of a desire to do their bit for the environment.

How is the directive to 'go vegan if you want to help save the planet' heard by a mother with two small children barely staying above the poverty line? It sounds like environmental activism via food is for the wealthy only, and that she has no recourse to care.

Michael Carolan, who teaches at Colorado State University, and is a visiting researcher at New Zealand's Auckland University and Australia's Queensland University of Technology, told me it can be harmful to identity and self-worth if people are unable to live up to the aspirations of middle-class consumers who are trying to eat environmentally and ethically. If the mum of two decides to spend a couple of dollars buying a McDonald's cheeseburger or a bag of crisps for their child, as a treat to make them feel special and loved, rather than devote those dollars to buying plant-based meat, it doesn't signal that she's unconcerned about the impact food has on the environment. It signals her love for her child.

In New Zealand, climate change has been branded by Prime Minister Jacinda Ardern as 'my generation's nuclear free-moment'.[44] We will be remembered by generations to come and split into camps: those who acted to curb emissions and those who obstructed to protect profit. The American economist Milton Friedman posited an oft-quoted theory that the market is like a vast democracy. So, every dollar is equivalent to a vote. An easy and convenient course being offered to consumers is to simply shop our way to fewer food-related greenhouse gas emissions by buying vegan products. But this act, as part of a broader democratic landscape, means some people get more votes because they have more dollars. 'There's a whole lot of risks going along with falling into that trap of focusing too much on consumption as the lever for change,' says Professor Carolan. 'Not just because it isn't a very good lever, but it also can result in some disingenuous social consequences, from shaming to people feeling alienated.'[45]

In Western secular culture, we're taught about the journey of man coming up through the ages as a hunter-gatherer who turns into a farmer. Protein was often sourced from the hunt, and it helped us to flourish physically and cognitively. But what if that's just a story we tell ourselves — the bit about needing to hunt and kill to be healthy?

I put this question to Dr Anne Louise Heath, who is an associate professor of nutrition at the University of Otago, New Zealand. Dr Heath's research is in maternal health, specifically iron deficiency in mothers and infants. She tells me we have a biased view when we look back on prehistory because anthropologists and archaeologists were, in the main, men, and so 'they found hunting a lot more exciting than, you know, four hours of collecting and gathering'.[46]

But in her view, the bulk of our protein and nutrient intake was the responsibility of early woman, not man. Plants, roots, seeds, small animals, and grubs were food for the family; the men would provide the feast, not the staple.[47]

Dr Heath is a sage guide as I contemplate a diet with very little meat. Unlike many in the West, I'm more interested in where my next (micronutrient) iron not (macronutrient) protein fix is coming from, because I have suffered from iron deficiency since I was a teenager, and it was exacerbated during my failed pregnancy.

This is hardly rare. When women start to menstruate, they lose blood each month, and logic dictates that the body needs to replenish those reserves. That requires iron, among other things. We humans absorb iron via food or supplements, mostly. And people absorb it differently. Some can satisfy their body's needs via a plant-based diet using the non-heme iron in plants, but others find red meat to be a more efficient carrier of the essential element heme iron. And essential it is.

Iron is part of haemoglobin, which is the molecule that carries oxygen from your lungs out to your tissues where it's needed. So that crippling lethargy I felt post miscarriage — that was my body not getting the oxygen it needed because I didn't have the iron reserves to make it all happen.

When I started reducing my red meat intake, I knew I risked descending into iron deficiency and anaemia again.[48] To operate on half battery is to miss the full menagerie of life's experiences, including being able to grow a healthy other life. Dr Heath, who has long followed a vegetarian diet, assured me that it was possible though. 'There is no such thing as vitamin beef,' she said, before counselling that I would need to think about three things.

One: make sure I maintain high iron intakes by consuming loads of legumes — lentils, chickpeas, hummus, falafel, and the like.

Two: avoid caffeine at mealtimes and have a good source of vitamin C like oranges or tomatoes to help absorb the iron from the aforementioned.[49]

And three: buy cast-iron cookware.

'Say what?'

Yes. Cook with cast iron. 'It's magic,' says Dr Heath. 'It's just marvellous.'

The iron, apparently, leaches into the food as you cook. This is the odd kind of factoid that I think must have been passed down among women once upon a time. But the bonds that bind my women are brittle, and the information about life in the kitchen has been lost in the race towards career success and urban convenience.

Dr Heath has researched the effects of iron deficiency for decades, but she tells me it's hard to quantify how prevalent the condition is among Australian and New Zealand women in any given year. Our systems are always in flux, with menstruation, pregnancy, stress, and dietary changes contributing to a collective tidal effect. One month we will be low, the next high.

But nutrition, or lack of, does play a role, and that means deficiencies can be tied to privilege, or rather, to underprivileged sectors of our society. Those, like me, who can afford to choose a balanced or on-the-go diet are likely able to access supplements when needed. But some just have to make do with what they can afford: white bread and eggs is a family dinner, or a meat pie cut in two.[50]

The number of people experiencing food insecurity may be as high as 40 per cent in New Zealand. In Australia, it's close to 13 per cent for the general population but double that for Indigenous and Torres Strait Islander peoples.[51] To be food secure is closely tied to disposable income, and the ability to acquire food that is nutritionally adequate, safe, and culturally nourishing.[52] To be insecure is to never have enough to cover all household expenses, and when that happens, it is women who are more likely to go without. Especially women who are single parents. When food is both sustenance and a symbol of care, it falls on the caregiver to provide.

There is an abundance of food to be had in New Zealand and Australia. Yet the vast majority of farming operations are growing commodities — beef, chickpeas, barley, wheat, peas — that are exported to larger markets where consumers are willing to pay a premium for food grown in developed, regulated countries.

For domestic consumers, the locally grown produce that makes its way into the food system, mostly via the supermarket duopoly, is often unaffordable once on the shelf. Half a kilo of newly dug Agria potatoes is far more expensive than a bag of crisps; a four-pack of two-minute noodles is cheaper than the sum of a packet of spaghetti, a tin of tomatoes, onion, garlic, carrot, celery, and beef mince — the ingredients for bolognese. So many have no choice but to reach for cheap imported processed foods to get by.

If plant-based products command a premium, and fresh whole foods are out of reach for many, how can we crawl our way out of global warming via food choices? The entire food system is geared to profit, not collective wellbeing.

In mid-January, Pat and I rise with the sun, hitch the trailer to the farm truck and drive to the flat paddocks at the top of the hill behind the house. The rye grass has gone to seed and turned golden in the heat. The long blades of rye, and the flower heads of clover and common yarrow, have a mustard tinge to the ends. They're burning off. The hay contractor has been, and 300 small rectangle hay bales lie scattered along tractor lines.

Soon we're joined by my sister and her family, and the day begins. Me at the wheel, my sister, husband, and brother-in-law jog-walking behind, heaving bales onto the trailer. My two-year-old nephew races between the bales, his jumpsuit flap flicking out behind him like a superhero's cape.

In this part of the world, hay making is at the behest of Tāwhirimātea, New Zealand's mythical weather god. You need a window — one day to cut, another to turn, another to bale and stack. All without a drop of rain, ideally. If you don't get it right, mouldy hay and hungry stock will follow. A wasted grass crop worth thousands of dollars.

The harsh sunlight often blinds Pat, obscuring a way beyond the task of the day at hand. He is too present; I am too often elsewhere, plotting a way out or at least forward. The light is a difficulty that drives me forward towards something I've yet to pinpoint.

I'm eight months pregnant, and this is the end of the work year. I've been climbing yard rails, mustering, rearing calves, and cleaning feed equipment every day since July, and I'm tired. I'm consuming three iron-supplement pills each day to keep my reserves up. The work of the year has made me physically strong; the food I grow and eat has kept me going. But the weight of obligation — to act as intermediary between my husband and father, and to something bigger and intangible — has created a stoop.

As I look out over the hay paddock, I picture instead rows of quinoa or chickpeas — but beyond are steeper hectares where no cropping tractor can roll. It's fit for grazing or being left alone to let the manuka trees take over once again. How will I know if I'm doing enough?

I'm thinking about what William the young vegan protestor said: *fight for what is just and right.* But how do you break and remake an entire farm? An entire food system? How do you know what's right?

3.

Trust in waste

We would often rise early in Sydney and drive to the edge of the city, where the Tasman Sea meets the city's famous beaches. It was the closest I got to my birth country for most of the year. We'd park in the Clovelly Beach carpark and run the ribbon track along the towering sandstone coastline and down the cascading steps towards Bondi. A montage of ritual practices spotted the beach: a couple sleeping entwined as the tide encroached, muscled football players running sprints up the sand, two women dressed in G-string bikinis extending a leg to the sky in a yoga offering. Pat stumbles a little.

Usually, we'd be early enough to see the sun rise. Afterwards, we'd drive to a tiny cafe in Clovelly where I'd order a hard-boiled egg with a side of feta and smashed avocado; Pat, a ham and cheese jaffle. The egg was creamy, a deep yellow that softened on the tongue as the accompanying sea salt popped as a top note. And we'd drink coffee. Sweaty, red-faced, and hungry, we'd sit quietly together and eat, with no thought to how this delicious food came to be on our plates.

We sacrificed nothing and wanted for nothing.

Sydney is a city that luxuriates in dining and eating experiences and trends, with an emphasis on trends. A couple of years before we left Sydney, a beloved pizzeria in our neighbouring suburb of Newtown switched to making vegan pizza, and people started queuing. They came because they wanted to sit in front of a tiered stack of red, moreish pizza and feel they were making the right food choice for the planet. The restaurant was applauded for its ethical and sustainable approach to pizza just because cheese was off the menu.

To prepare a vegan pizza is relatively simple: leave off the meat and replace the cheese with something made from nuts. But to cook vegan food well, in a way that satisfies both nutritional and emotional needs, is complicated — especially if you're following food trends rather than tapping into old food traditions.

In my final weeks of pregnancy, I start to experiment with everything other than meat: a dense mix of quinoa, beans, toasted nuts, garlic, and onion, seasoned with umami-rich dried mushroom and seaweed spice, gives a hearty approximation of a burger patty. A satisfying, slow broth made from brown rice miso paste, charred onion, garlic and ginger, carrots, leek, and spices passes for a bowl of pho when served with rice noodles and herbs.

I use taste to guide me. Trying my concoctions as they brew, aiming for a heartiness I have long thought only animal products could provide. These experiments give me an odd sort of power in the kitchen, a space I'd long along rejected in a flippant, ill-conceived feminist gesture, believing it was a place to use but not be tied to.

Yet this type of cooking — from scratch — takes time, process, and knowledge. Three things an increasing number of urban people lack. Australia may have long legislated a 38-hour working week, but in the past decade a 50-hour work week, or juggling multiple part time gigs, has become common, according to the Australian Institute of Health and Welfare. Little wonder fast and convenient food is sought.[1] Less wonder still that time-poor people are following the latest trend to shop their way out of climate change.

Many of us are learning and updating our food habits from Instagram chefs and influencers, and cookbook authors who've made a career diluting the flavours and textures of a cuisine not their own for the sake of a mostly white middle-class readership: people like me, in truth, who gravitate to the flavours and textures discovered in Sydney's food courts and neighbourhood eateries. We buy ingredients for one-time recipes, and leave unwanted items to pile and waste in the fridge. At the end of the month or year, the clean-out begins, and the waste piles up, ready for landfill.

Most Australians throw out around 300 kilograms of food every year. The methane from food waste generates the equivalent of almost 7 million tonnes of carbon dioxide each year.[2] Some 12 per cent of New Zealand's anthropogenic methane emissions — that which are caused by human activity — also come from food waste. *One in three kilograms of food produced is wasted* — it's rejected by suppliers, left to rot in paddocks, or dumped.

Globally, the combined greenhouse gas emissions from food waste alone are so significant that it would be ranked as the third-largest polluter if considered a country, behind China and the United States. Food waste is as large a problem to solve as biogenic methane emissions from cattle, due to its presence in landfill, where it becomes a source of anthropogenic methane gas. Yet the call to cut beef and dairy dominates the change-maker conversation, and big food and big agriculture capitalise by creating new, so-called planet-friendly foodstuffs, when instead we should be resolutely focused on stopping the wasteful practices embedded in the current food system.

Why is that? Is it because we're waiting for someone to tell us how to cook and shop before we act?

I know I was. Because honestly, who has time to figure out the best course? It's far easier to follow the headlines: 'Our addiction to big beef has ruined the planet!' And easier to wait for someone to tell us what to do rather than look to ourselves for answers that may result in something we're not ready for: real change. Wholesale change. A sacrifice, of sorts.

At BuzzFeed Australia, where I worked as Managing Editor, a change-the-world influencer energy was palpable. We were a media organisation that catered for diversity, and that was a bit revolutionary — at least, until the redundancies started, and the business constricted in cost-saving measures before the company's public listing on the US NASDAQ.

But for a while, a workday was dedicated to wide-eyed experimentation and iteration, a core ethos of the start-up world. The intent was to create, release, and learn. And do it all over again until, if you were a video creator or a culture writer, you hit a million views or 25 million views. And nowhere was this more evident than in the *Tasty* kitchens.

Tasty is BuzzFeed's successful food-content offshoot, a brand that popularised the birds-eye cooking demonstration videos that initially privileged stretchy cheese, party foods, Milo pudding made in the microwave, and nostalgic cooking like cheese dip served in a hollowed-out bread loaf — perfect for 'game day'. At its height, *Tasty* and the hundreds of producers and assistants who made food videos in purpose-built kitchens in Los Angeles, New York, Tokyo, Sydney, and London, would capture the imagination and time of tens of millions of viewers.

The company likes to quote an example of a video about a kitchen appliance, the Oster Grill, to prove its ability to translate social media views into consumer action and client revenue. The video was 60 seconds long and featured a pair of hands making a jalapeño and cheese-stuffed beef hamburger. The weekend it was released it generated 20 million views, and as a result, the client asked BuzzFeed to pull the other Oster Grill videos from its publishing schedule. They'd sold out. There were no more Oster Grills to be had. Stop the videos!

The ability to persuade is paramount when one is seeking capital to push a start-up along. Historically, data and revenue and production projections have been prioritised, but increasingly the questions many investors ask of start-ups are these: are the numbers real? Is the demand for your product really there? Or is it just hype … slick marketing?

The questions many consumers are increasingly asking those same companies are: can we trust what you're selling me is what you say it is? Will the grill make the cheese in my beef burger melt into decadent stringy ribbons; if I swap the beef burger for a plant-based burger, will I be helping the environment?

Knowing who to believe and trust when it comes to climate change information is fraught. In 2020, the Reuters Institute released its 'Digital News Report' with data sourced from a survey of more than 80,000 people in 40 markets. The report's insights on how media consumers are accessing information about climate change illustrate how easy it is for nefarious players to dominate the conversation, and how simple it is to sway readers into consumer habits. The report found that 69 per cent of respondents believe climate change is an extremely or very serious issue.[3] And while most people over 35 still source most of their information on this issue via television, Generation Z — those born between 1997 and the early 2010s — are looking to alternative sources of media for their information: Instagram and activist blogs, mostly.

One 19-year-old woman from the United Kingdom said: 'I follow climate activists like Greta Thunberg [on social media] and follow what she does.' Another, a young woman in her twenties from the US, prefers to follow Hollywood actors and Australian models for climate change information: 'Leonardo DiCaprio posts on Instagram about climate change a lot – so does Cody Simpson.'

In November 2019, Leonardo DiCaprio posted a photo of himself and Swedish climate activist Greta Thunberg on Instagram. It read: '... History will judge us for what we do today to help guarantee that future generations can enjoy the same liveable planet that we so clearly have taken for granted ...' Liked more than 4.5 million times, the post goes on to plead for world leaders to act.

Meanwhile, at the other end of the age spectrum, some 8 per cent of Australian respondents and 12 per cent in the United States, all in the older demographic, don't believe climate change is a serious issue at all. These people are just like my dad.

The polar conversations I have with my boomer father and vegan-conscious niece aren't unusual. They're being played out the world over as climate change consensus gives way to partisanship, scare-mongering, and the well-intentioned offerings of celebrity activists.

In other words, division.

In October 2020, Greta Thunberg was raging at European politicians. She released a message to her 10 million Instagram followers stating: 'The EU is cheating with numbers — and stealing our future. Since no one else seemed to care, we had to do the research ourselves. Our future is after all at stake.' Alongside a photo of Thunberg and three other young COVID-masked female climate activists, the message targeted members of the European Parliament who approved reforms to the Common Agricultural Policy (CAP) that critics say pit farmers against environmentalists.

On the reform agenda, among other issues, was a push to cut subsidies for factory farming, and to increase landscape protections for grass and peatlands to create safe habitats for at-risk species. Neither made it through the vote. In an open letter collecting signatures, Thunberg blasted the decision, accusing the members of prioritising profit and economic greed over the planet's habitability.

Since the 1990s, the primary recipients of the European Union's

CAP subsidy have been multinational companies like Nestlé and Tate & Lyle, not smallholder farmers.[4] And a hot topic during the policy debate was the issue of food labelling in the European Union.[5] Beef industry advocates were pushing to restrict plant-based food manufacturers from using common terms like 'burger', 'steak', and 'sausage' on non-meat products, calling for the terms to be 'reserved exclusively for products containing meat'.

Thunberg told her followers that this niche issue obfuscates the bigger challenges, tweeting: 'While media was reporting on "names of vegan hot dogs" the EU parliament signed away €387bn [£350bn] to a new agricultural policy that basically means surrender on climate & environment.'

Yet I'm convinced that this labelling battle goes to the heart of agricultural change in the European Union and elsewhere, because it is one of the few ways in which vested interests can be revealed. Over the years, product labels have been regulatory battlegrounds where public health experts, nutritionists, agriculture lobbyists, and trade negotiators vie for territory. Nutritional data, country of origin, allergen information, quality control marks, 'made in' brand marks all have value on labels. Now environmental impacts are being added via certification and compliance logos. Labelling can cut through the hype, but only if it's a label that stands for transparency and truth, and not greenwashing.

The National Party of Australia, a governing coalition partner with the Liberals, declared food labelling to be the 'number one issue' in food regulation in 2019.[6] Former agriculture minister Bridget McKenzie said labelling was mostly a health issue. She told the party's council members that the Nationals' stance was an objection to meat and dairy protein replacements that were 'trading off the health and nutritional benefits of milk and milk products'.[7]

In the United States, Elizabeth Derbes has been monitoring the wave of legislative bills being presented at state government level aimed at limiting the use of meat terms in labelling. An attorney for the Good Food Institute, a non-profit group advocating on behalf of plant-based and cultivated-meat manufacturers, Elizabeth told me that the push has come from cattle-rancher associations who are fearful consumers will move away from beef grown on the range or in the feedlots towards plant-based products or cultivated meat, clean meat — i.e., meat grown from cells in the lab. It's not yet on the market at scale, but the push is on to make the products accessible at a price point that's affordable. Early attempts priced a burger at more than US $300,000 but Just Inc, a company that makes egg from mung beans, has managed to get its production price to under $50 per nugget.[8] And in December 2020, it announced that its cultivated-chicken nuggets, which comprise 70 per cent lab-grown meat and 30 per cent plant-based filler, had been approved for sale in Singapore.[9]

Yet, as Tom Philpott from *Mother Jones* writes, the start-ups developing lab-grown meats have promised imminent mass production for almost 15 years, and there is little affordable product on sale, despite the venture-capitalist money pouring in.[10] So you may wonder why the effort, why all the research and development, and why all the legal and legislative wrangling from meat and rancher associations? Money, of course. The clean meat sector wants a slice of the US$1.4 trillion global meat industry. Cattle ranchers want to protect terms like 'steak' and 'hamburger' and, like their European counterparts, to mark it out as food that comes from a real animal via a slaughterhouse, because they don't want to share the pie.

When I spoke with Derbes, cookie-cutter bills were in the legislative process in 25 US states, and she and her colleagues were tracking them all, lobbying for the clean and plant-based industry at every turn.

Dan Altschuler Malek, a San Francisco–based managing partner at New Crop Capital, a fund *The Guardian* describes as 'a vegan venture fund fighting for animal rights',[11] explains to me one day over a crackly Skype line that the fund, on behalf of its clients, is trying to disrupt the animal protein industry because it thinks 'there is a better future' that mostly involves alternative proteins.

They've identified a number of food tech start-ups specialising in vegan products including Abbot's Butcher, Beyond Meats, Black Sheep Foods, Good Catch, Miyoko's Creamery, and SunFed, one of the few New Zealand companies in the New Crop portfolio. Altschuler Malek says the measure of a good investment in the food space is simply taste. They're not looking at profit, yet — 'that comes later' — but at whether the product on offer actually tastes any good. And then whether it can scale to become the disruptor New Crop's investors expect.

I asked Altschuler Malek whether he thought consumers had a right to know whose money was funding their latest plant-based purchase, especially those carrying a vegan label. What if the principal funder was a big beef company, like Tyson Foods? This is a company that until 2019 had a sizeable share position in Beyond Meats, and now has its own plant-based products offered under the label 'Root & Raised', which makes protein nuggets and blended Angus beef and pea-protein burgers, among other things.

He was befuddled by the question: 'I don't see the relationship,' he said, before elaborating that he couldn't think of any case where a consumer would be actively looking at all the investors in a company to feel good about their purchases. That's just not how people shop, apparently. And yet, increasingly, it is.

Innova, a market research firm, found that transparency was the top consumer trend in 2021.[12] Consumers say they'll switch food brands for a product with more information about how and where the food was made.[13] So, in response, the French company Connecting Food has devised blockchain technology to create an audit system for food producer clients keen to present full supply chain information to shoppers. And in a similar vein, CoGo, an app developed by charity boss turned tech developer Ben Gleisner, can match a consumer's values with a company that is working to lessen its impact on the planet. That relationship Altschuler Malek couldn't quite see is manifesting, whether investors like it or not. We need this transparency to go mainstream and make it onto the product label. Without supply chain transparency, we're all duped. You won't know if your favourite green-tick plant-based or vegan brand is owned, funded, or backed by the largest producer of beef in the United States, or the world's largest meat producer which has also been linked to deforestation projects in the Amazon,[14] unless transparency becomes something tangible on a label and not just a cute app.

These companies are in the business of making protein and selling it. Where it comes from is of less concern than the bottom line, and so the food system has a trust problem. It's so hard to know who to believe. Who will guide us towards a cooler future? Is it Greta? Is it Nestlé?

Or is it me?

With the smell of jasmine hanging in the evening air, we bring our baby daughter home from the hospital. I've been away from the farm for a week, slowly rebuilding, with the aid of multiple iron infusions, after losing 1.6 litres of blood in the seconds after birth; a birth that started on Friday morning and didn't end until 6:23am on Sunday.

My first meal at home is seared sirloin steak sliced thinly and layered atop bread, salad greens, and hummus. A metallic clang hangs in my mouth after each mouthful. And I go back for seconds, but of just the beef. I want to suck the bloody juices off my plate, but I'm holding my baby girl. Who has fallen asleep.

I am eating meat again. Voraciously. Listening to my body, not the chorus of angry voices. I eat and stare down the headline that's become truth over the past two years: to eat meat is to degrade the planet. The statement is not wrong, but nor is it nuanced. My gut tells me there is such a thing as a meat-eating environmentalist, but my head tells me I'm on my own. Around me, people have chosen one or the other as a political statement.

Is it my role to build a bridge between the two? To fix the trust issue at the heart of the food system? High on postpartum endorphins, I say YES! I can do something that will give the critics a reason to trust cattle farmers again, and prove people can eat meat and help heal the planet.

Before the heat of the day builds, a truck pulls into the farm shed compound, driven by a butcher and his assistant. We've arranged to have a two-year-old cattle beast killed to fill our freezer, and to satisfy my cravings. They park and unload a rifle case. A Jersey-Hereford heifer weighing around 500 kilograms draws its head up sharply, pivoting toward them as they creep like hunters towards the yard fence.

I am standing at the top of the hill above the sheds with our baby asleep in a covered buggy watching the scene unfold. It's a hunt like no other: the assistant waves his arms above his head, drawing the heifer's attention, and then the beast drops dead. Quickly. Seconds between life and death.

Soon, it is hoisted by the back legs over the home-kill truck crane. The animal rests splayed, two metres up. It's sliced open, the huge gut network removed. The skin is removed. The head is removed. The carcass is packed into the truck and leaves with the butcher. In two weeks, we'll collect more than 200 kilograms of meat. But the waste — the offal, head, skin, and blood — is for me. Pat scoops it into the tractor bucket and drops it from aloft into a hole I've dug behind our vegetable garden. Beneath is a layer of small sticks; over it go leaves, manure, then dirt and a layer of mesh that I fix with pegs to stop stray animals from digging at the decomposing mess. The waste will become fuel: fertile soil for the vegetable garden.

I've been thinking a lot about waste since we killed that heifer for meat. Waste and its myriad definitions. Our Friesian bull calves have long been considered a waste product of the dairy industry because they can never produce milk. But that same industry has, since the 1970s, been a global leader in up-cycling waste made during the production of milk powder, notably whey, which is now one of the most profitable protein ingredients produced worldwide.[15]

My suspicion of plant-based manufacturing was tempered when I discovered that many of the raw materials that can be used to produce plant proteins are also 'by-products'. Oil seeds, for example, produce waste that is high in protein and fibre. Garden peas, when produced to extract protein, also produce a food-grade starch, plus an insoluble fibre that could be used as a prebiotic food ingredient or biomaterial for reinforcing bioplastics. All from a little garden pea.[16]

Waste shouldn't be seen as something unwanted. It can provide an opportunity to break food hierarchies. But that's an opportunity disconnected from the notion — so regularly perpetuated in the conversation about climate change — that an individual can effect change at the point of consumption. This 'shop to change' idea assumes that all individuals are avid consumers. Yet if you're relying on a foodbank parcel each week, or you grow your own food, by this logic, your ability to act for the common good is neutered, regardless of your will.

Remember Milton Friedman's theory of the market as democracy? Every time you enter the store you vote your way to creating change. But what if the change you want is not on the shelf? If you shop for food in a supermarket in Australia and New Zealand, you are less likely to see a range of 'change options' than anywhere else in the Western world, for the simple reason that two food companies — Coles and Woolworths Group in Australia, FoodStuffs and Woolworths Group in New Zealand — control an 80 per cent share of retail food sales.[17]

Shopping has its inequities, but the commitment to not waste food is an equaliser: in the home, and, it seems, in the manufacturing line.

Make every morsel count.

It can be that simple.

Soon after our baby was born, I met Sofia DeKovic and Jennifer Long — a Chilean and a Scot who have built a business around waste. The pair sell weekly produce boxes made up of vegetables rejected by supermarkets and exporters. They market it as 'ugly' food, but when I started subscribing, it was clear there was very little wrong with the produce. Often it was greens that had bolted too early because the weather had been warm and wet, and the produce was ready before the contracted date. More still was too big, or too small, for the supermarket's aesthetic standard.

It's easy enough to say, 'just don't waste food' or 'don't buy what you don't need', but the waste problem and our ability to combat it is in part about contracts, and nature's inability to understand or appreciate the nuances of contract law. If farmers can't meet the aesthetic standards of a produce contract, or meet the harvest delivery window, then that food is left to rot or sold as animal food. It has little or no value.

The farmer shoulders that risk and vows to do better next season, to be more precise. There is little generosity from retailers, who for the most part choose not to explain to customers that fresh food isn't supposed to be perfect; it's small, misshapen, sun-burnt, and hail-marked.

Countdown, a subsidiary of Woolworths Group, has publicly

pledged to send zero food waste to landfill from stores by 2025. To do that, the company is working on its 'forecasting'. Kiri Hannifin, Countdown's General Manager Sustainability, tells me the aim is 'to sell everything that we buy in, and to only stock what we know we will sell'.[18] That, she says, has reduced stock loss by 29 per cent over the past five years. The publicly listed company also works with Ecostock to turn unsold food into animal feed. And supermarkets across Australia and New Zealand are prioritising rescue programs in collaboration with food banks and social services, which lands us back at food inequity again too soon — another problem to solve alongside climate change. But for now, let's focus on carrots.

One carrot farmer who sells to Jennifer and Sophia had around 40 per cent of an annual crop rejected because the carrots were too long or slightly crooked. An avocado orchardist was offered eight cents a kilo for cattle to eat their sun-burnt export crop; Jennifer and Sophia paid a dollar per fruit instead. There was nothing wrong with the sweet creamy flesh inside — discoloured skin had caused the rejection.

Jennifer and Sophia are determined to pay a fair price, to make it more worthwhile for farmers to sell to them than to dump their food. As winter approaches, I strike a deal: what's not sold to their customers comes to us. The waste of the waste is now cattle food: our yearling heifers feast on Brussels sprouts, carrots, parsnips, and swedes. It's a small thing, but small things add up. The food isn't dumped, at least. And we start to close the circle around our region, start to drop our carbon footprint. Food grown here is eaten here.

Professor Carolan reminded me that there is no such thing as eating food that has no impact on some aspect of the world. We can't solve the problem of climate change and environmental degradation by reimagining the food system without ignoring or destroying something else. 'People have to really reflect upon themselves about what it is that they value,' he told me. And then pick something. Because to say we value everything can lead to total paralysis.[19]

Pick something.

If Pat and I pick the environment over and above all else — specifically, do all we can to drop our greenhouse gas footprint — then I ignore all the other inequity issues within the food system. And I risk sacrificing my relationship with my disbelieving dad, who has picked the status quo. Is that a sacrifice worth making? Family and the collective for the environment?

As I stand outside the slider doors of the farmhouse, watching the hawks work and the sun rays flicker off the surface of the little pond below the house, I realise that the prospect of sacrifice stings like a collective cut for a group of people whose identity is wedded to the status quo.

My dad and uncle are farmers. They are part of a generation whose collective viewpoints, mores, and traditions form an identity that screams FARMER. To ask them to give that up is akin to asking them to volunteer for an amputation. Their views on the scale of global warming — and the farming and food sector's contribution to climate change — frustrate me, and I've asked this collective over and over again to look at alternatives. To change. To sacrifice their way of doing things for something different. But they barely budge.

Farming is the backbone of the economy, they say. It shouldn't change.

There's always been herds of ruminants on the planet, that's not what's causing global warming.

We need to crop each year, or the cattle will starve, many say.

We need to use tonnes of fertiliser, or the grass won't grow.

We need to export the bulk of our produce, or someone else will.

We need to sell to overseas markets not to locals, because that's where the money is, and the economy needs the revenue earned from exports.

We need to push the land as far as it can go … what else is it good for?

'I want to talk to you and Dad together,' I tell my uncle when I call to invite him over. 'I want to find out if we can talk about all this stuff without arguing.'

'Oh well, I always disagree with him,' says my uncle, who rose to the top of large corporate farming entities, of my dad.

'What do you do?' I ask.

'I just walk out.'

I'll have food ready. Laid out on the table so we can share something before we start talking. 'Come round Tuesday afternoon, and we'll see how it goes.'

I speak with my husband's niece too, to ask what she'd consider giving up. She's in her final year of primary school in Sydney, and she has high school on her mind. As we chat, she sits with her mum and sister on the couch, eating fruit and yoghurt. She tells me her friends at school are mostly concerned about 'social justice' issues now. Climate change was more important last summer.

'People were very focused then,' she says. Still, I ask if her friends have continued to talk about giving up meat as a climate solution, whether they still believe the agriculture sector is to blame for our collective predicament. The update, she says, is that more people at her school are vegetarian, and increasingly, flexitarian. 'Is that the right word?' she asks.

Yes, I say. That's the right word.

It means reduce.

All of us have a responsibility for fixing warming, she tells me.

'You seem to have a fairly solid grasp of the key points,' I remark. 'Is it taught in school?'

'No, not really,' she says with a sigh. 'It's just that we're aware.'

Aware.

In thinking about the nature of conflict, American writer Sarah Schulman, who describes herself as an 'active participant citizen', reasons that it is the community surrounding a conflict that is the source of its resolution. If the conflict is cattle and climate change, and the community is farming, then I think I may be central to the resolution through the act of not overreacting, and a commitment to not quit. If I create space where alternatives of understanding could be aired without shame or shunning, if I'm patient with my dad and uncle, can I guide them forward?

They know the land is changing. It's a fact of nature that they can see with their own eyes, and feel in the sweat that clusters on the brow when a chill should be causing them to hunch their shoulders and dip their head to the wind. The seasons have elongated. The opening weekend of duck-shooting season in May is no longer cold. Winter was once 100 days, a season when grass would not grow. The summer now seemingly starts in September, three months early, and runs through to March.

But there's always been warming on the planet. Farming isn't to blame, they say. 'What about the millions of buffalo gone,' they say. Methane has always been emitted, it's just now cows rather than buffalo herds doing the burping. Farming isn't to blame.

I have food ready when my father and uncle arrive on this Tuesday afternoon. Hours earlier I've combined tepid water with yeast and mixed through 90 grams of potato bread starter. The beginnings of local bread, a recipe I've learned from chef Monique Fiso. Now I sift in flour with toasted fennel and cumin seeds and grated sweet potato, combine the ingredients, and then leave it alone to rise.

I process chickpeas, tahini, garlic, lemon juice, and salt in a food processor and then dribble ice-cold water into the blend as the blades work the paste in circles. A hummus recipe I've learned from the teachings of chef Samrin Nosrat, who says, 'Don't add olive oil

85

until you serve, and then it's just a garnish.' My hummus transitions from a lumpy mess to velvet-smooth paste.

I pan-fry a small portion of beef mince, just 42 grams — three portions of the flexitarian diet recommendation for daily red meat — with garlic, onions, red peppers, paprika, cumin, salt, and pepper, until knobbly and crisp. I add parsley. A recipe I've made up to make the most of leftovers and garden herbs.

I shape the bread into flat rounds and cook each one on an iron hotplate drizzled with butter, five minutes each side. I serve the hummus topped with spiced beef next to the piping hot bread and offer the plates up to my dad and uncle. They take small spoonfuls, tasting first, just a tiny bite before going for a larger one. 'I'd love to be able to do things like this,' my uncle says. 'I'm just not into recipes, not into mixing things up.' The food is peculiar to them, unidentifiable, but it's eaten, and savoured in small mouthfuls. Flavour suspends judgement for just a few minutes. And we talk.

I think they can be swayed from their spot. We pass plates as my uncle offers that one of his granddaughters 'eats like you'. She's 15, he tells me, and 'really aware of her food'.

Aware. Just like my husband's niece.

My uncle inches towards the bridge I'm trying to build, thinking of his grandchildren. And as lunch draws to an end, he stops making jokes. And says he wants to keep talking. They leave food on the plates. There's a lot left over.

If I've learned anything in the past few years, it's that change starts with a simple shift in personal behaviour, and that gesture can make a huge difference in collective experiences.[20] I was waiting for them to shift towards me on the assumption that I knew what was good for us all. That their change would give us all a future. But perhaps that's not it.

Pat didn't join us for lunch; he can't bear to listen to the obfuscation that's passed off as farming knowledge. I make him a plate from the untouched food. He returns once they've left, and I watch him eat, thoughtful and ravenous. I look at him, nothing beyond. And in that moment, I stop making space for my dad. 'Just believe in me,' I hear him think. 'Not them.'

I can compromise my values so a generation of farmers feel comfortable enough to talk, think, and capture a passing remark that brings them forward into a future which shakes their identity as 'farmers'. Or I can leave them where they stand, where they feel comfortable, where their accumulated knowledge makes sense and is valued. While I've been trying to change others' views and figure out how shopping for food can create change, Pat has been quietly creating a new world. And I didn't even notice.

That's how change starts. Small habits changed. Big thinking applied to daily problems.

At some point, it had occurred to Pat that we, as farmers, are as much consumers as the people in the supermarkets shopping their way towards a cooler planet. And in turn, so too are supermarkets and big food manufacturers consumers, via their procurement policies for manufacturing and stocking food. If a different way of buying right through the supply chain was effected, we wouldn't all be so reliant on the consumer to 'shop for the planet', would we? The theory that every dollar is equivalent to a vote becomes all the more powerful in Pat's eyes, because the democratic landscape is extended to strategy documents, corporate leaders, and individuals working throughout the corporate supply chain.

It didn't take long to learn that conventional farming is dominated by monopoly players. It takes shape like an hourglass. The base is family-run farms, millions of them. At the top are millions of consumers. But every point along the funnel is owned by just a small handful of multinational companies. Each country varies, but farm inputs — seeds, fertilisers, machinery manufacturers — generally come from between five to ten companies.[21]

The product is processed by just a handful of companies that then funnel the food to export markets or domestic wholesalers, and then on to one of two retailers. It's a very small pool of players that have amassed a significant amount of wealth and power.[22]

As summer faded, Pat made a list. It detailed the biggest purchases the farm makes each year, and he vowed to cross each one off before the year end.

NZ$6,000 on fertiliser and spreading.

NZ$82,000 on milk powder to feed the calves.

NZ$15,000 on grains and meal to feed the calves before they eat grass.

NZ$12,000 for a contractor to cut silage and hay.

NZ$8,000 on diesel and petrol — fossil fuels.

What does farming look like if we stop fuelling the funnel? If we stop paying everyone who tells us we must?

Food systems researcher Laura Pereira writes that transformative change often occurs when a window of opportunity opens up, and innovations that have been relatively marginal can suddenly be taken up and set loose to shape a new set of processes.[23]

To bury offal to grow vegetables isn't a revolutionary innovation. But what if it scales? What if its core function — to fertilise the soil for the growing of food — ruptures a key element in the conventional farming supply chain: the use of synthetic fertiliser?

When I visited the young dairy farmers at their 1,000-hectare farm in Otago, they chatted excitedly about a contraption they'd purchased to affix to the tractor. It was a compost turner, designed to work the massive rows of organic matter that was stewing into compost. In the mounds were plant matter, bark chips, a dead cow or calf or two, and all manner of waste from the calf sheds and calving pad. It is their fertiliser — a marginal idea that's scaling. It's protest disguised as a compost row.

There is reward for living within a thriving ecosystem, one that sees death and life in succession. But it's not easy to decipher amid the labour and the toil. There's no instant gratification.

I came to this farm thinking I could go back at any time. To Sydney. To a way of life that for a while was draining but also nourished the part of me that is charged amid noise and the pulsing energy of a cityscape.

But it's gone now. Climate change doesn't celebrate the frivolity of easy holidays abroad, convenient food-court meals, and cooking on-trend then burying the waste of the endeavour. When I realised that escape route had disappeared, I felt a deep loss. I don't want to be entirely of my new community — among those farmers who are out of step with the way we see the future — but the old community, my Sydney life, has so changed.

This is what untethering feels like. Sadness laced with uncertainty as the rhythms of life shift. And a mass untethering is *exactly* what sustainable food advocates and experts are telling me we all need to do. Unhook from a global food regime that treats food as a commodity that can sustain imaginary communities, who care little about where their food comes from, how it was made, or where it ends up.

When I ask you what sacrifice you're willing to make, you can choose to not waste food — buy less more often, eat leftovers, make soups. Anything to avoid putting food that rots in the kitchen rubbish bin. And that will make a difference.

Or you can walk with us, away from the status quo, to reimagine the future — but that requires a reckoning with the past. These systems I'm trying to break haven't been conjured out of nothing; they are the product of deals brokered long ago when my family were landowners, and decisions in this part of world were made by men in gilded halls.

As our baby girl sleeps, Pat and I sit outside our farmhouse sliding door, overlooking a small body of water optimistically known as Lake Elliot. Our work, feeding calves, is done, and we watch the mallard ducks paddle around the surface, often taking flight, startled by the noise of a truck horn that echoes through the valley.

A towering wall of radiata pine trees once flanked the water on the north side, shielding the lake. They were felled soon after we arrived by the landowners. A mess of pine trunks, bark, and branches was left to rot on the water side.

A crew of men arrived in the weeks that followed to plant totura, flax, and hebe seedlings. And then the willow trees flanking the stream feeding the small lake were injected with poison. A helicopter arrived one autumn morning, flying low over the lake, weaving through the remaining trees, dropping a spray of poison onto the blackberry that clogged the flanks of the stream. More seedlings were planted. Around us, nature brought from the north is dying to give way to life that belongs here.

And we cheered the destruction.

4.

Corn fritters in
the colonies

There's a phrase that's used in Māoridom in Aotearoa with a certain reverence: *tangata whenua*. If someone is tangata whenua then they are 'of the land'.

The land is me and I am the land.

I am not tangata whenua.

There's another word I've been trying to understand: *Turangawaewae*. In words, it might mean a place for my feet to stand, but in life, it covers a connection someone has to a certain place and a responsibility to nurture that land and all that comes from it.

This was explained to me by a tall, gracious man named Murray Hemi. He towered over me when we first met at a cafe in the small town near where I farm. But he soon folded neatly into a chair, taking up little more room than me.

I'd emailed him earlier in the week because I was interested in learning more about his employer: a dairy company called Miraka, which is located about 15 minutes from our farm. The company had started to win awards for taking a sustainable approach to dairy production, and I figured in their strategy might be a clue to where we had gone so wrong so quickly.

Cold-calling someone to ask for advice never stops being nerve-racking. In my media life these conversations would be formal interviews and considered news gathering. But farming is reliant on knowledge being shared — or withheld if you're not in the right clique or from the right family. My challenge was to figure out whose knowledge to value and to work out what, if any, responsibility I had to the land we'd ended up on.

We needed to make money, and I'd had an idea that I wanted to produce good food, because it was food that had healed me when I was at my lowest. But I'm not of this land. And that fact is a big part of the reason why I left New Zealand at the age of 22 to live in big cities in other parts of the world — I thought my place was elsewhere. In time, though, I learned that the cities I chose to live in aren't neutral. I did not belong in Sydney on Gadigal Country any more than I did in London or on the plot of land where I now live in New Zealand. Land ownership has long been seen as the mark of belonging. But I'm starting to realise that belonging is something earned, not inherited or bought.

When Murray talks about farming, his remarks don't fit into the lineage of farm systems and land ownership that's recorded as history in this part of the world. Hundreds of years ago, Europe's industrial revolution fractured an agrarian society by sending workers into the city, but here, and in countries including Australia, Canada, and the United States, it forever changed the fabric of the land, because the colonies became Europe's food bowl.

The colonial-diasporic food regime ruptured the ceiling of innovation because there was no perceivable limit to growth.[1] If a ballooning population needed more food, food would be grown.

If land was needed to grow that food, it would be taken. If natural resources were needed to fuel factories, resources would be found.

This process was shunted into top gear in the twentieth century during the Green Revolution, aided by synthetic fertilisers designed to make crops and grass grow more. Suddenly, so much more could be produced. With intensification came the advent of pesticides, heavy machinery, and an increasing reliance on industrial inputs. The supermarket network that flourished in the twentieth century capitalised on this output, and food manufacturing locked into place. And even as the colonial era sank below the horizon, with the rise of globalisation the colonies kept pumping out foodstuff.[2] Commodities — beef, milk powder, barley, corn, soy, and the like — are sent into the global food manufacturing system and spat out onto the shelves of supermarkets the world over.

But that's a world away from Murray's version of farming.

His job title is 'Kaitiaki o te ara Miraka' which he tells me means, roughly, 'general manager of environmental leadership', but his role is that of guardian or protector. Not like a security guard, but more the person who considers how this Māori-owned company impacts the world in which it operates — from the environment to the community.

Translations are often imperfect; more so, if the word you're trying to translate encompasses a philosophical position that grounds a people. Like *kaitiakitanga* does for New Zealand's Māori. Murray is Ngāti Kahungunu. His iwi or tribe is from the southeast of New Zealand's North Island. When we first met, we talked about all sorts of things. Murray wanted to know who I was, where I was from, what my thoughts were on various topics. He leaned over a sandwich taking neat bites and talked. I listened and answered when asked. In the weeks that followed we'd talk more, and he patiently and generously unpacked a way of life that is innate to him and complicated to me. This is the labour often demanded from indigenous people — to educate for free those who don't understand. I sat, the assumed right to ask draped over my shoulders in a resplendent cloak of privilege. And I listened. For the phrase kaitiakitanga, he gave me this to work with: 'I guess a soft translation of that would be stewardship or guardianship for the natural environment.' But it's more than that, he says; it's about connection, balance, and long-term focus on the future of each generation.

And as we talked, and his generosity unfurled, he told me this responsibility is not limited to Māori. It might be an indigenous way of thinking, but in colonial countries there are families who have lived on the same farm or plot of land for seven or eight generations. Over a hundred years or more, the rhythms of nature have created weather lines on their familial skin, and they, too, have a responsibility to that place — more so if they choose to dwell on how they came to own that land.

The same can be said for a sixteenth-generation shepherd in the Cumbria hills of England or Castilla-La Mancha. These people have a responsibility to the land.

As do I.

But how we all choose to execute it is the story of modern farming and food production. Some will push the land to breaking point; others will guard and nurture it for generations to come. And that's where the problems and confusion start.

Is there a right and a wrong way to farm?

Can we farm as guardians to protect the planet and feed the world?

My ancestors — my mum's family — were in England, Ireland, India, then England, and India again. They made money, lost money, and made it again. And then the patriarch, in need of respite, bought a farm.

The mythologising required to create family legacies or national identities is a haphazard morphing of the actions we aspire to and the acts we commit. The same today as in 1881, when my ancestors arrived in New Zealand via Australia from India. Thomas, my mum's great-grandfather, was sometimes known as a nabob in this part of the world. A man who amassed wealth in India when it was under British rule. Had he returned to England rather than travelled south to New Zealand, he would not have been tagged with the term, reserved as it was for the powerbrokers of the East India Company who made their wealth by stripping the sub-continent of its assets and then returned to buy land and estates in England under the suspicious watch of the landed establishment. 'Those who owned land,' observes British Empire scholar Tillman Nechtman, 'had a stake in it and in the fate of the country.'[3] The land was the nation, and it was the strength of the nation's people. So, the wealthy wanted land. Power was cemented in the West's concept of individual ownership.

My ancestor Thomas was an engineer, not an East India Company man. He trained under Isambard Kingdom Brunel, and built bridges under the company name Thomas White & Co. The 'Co' were his sons — James, William, Thomas, Frank, Hugh, and Charles, who was known as Dick. To secure work in India, Thomas would write to the chief engineer of the London-headquartered Bombay, Baroda, and Central India Railway Company, one of the guaranteed railways originally authorised by the East India Company.

Bombay, 6th August 1877.
Sir,
We beg to offer to execute the Indian work of erection of the New Nurbudda Bridge, according to Sir John Hawkshaw's Specifications and Plan, at the rates and for the

sums as per tender attached; also to find the required security, and to carry on the works as rapidly as the delivery of the Iron will allow (which, if delivered in sufficient quantity to enable about 17 columns or one-third of the Bridge being pitched by the end of the year, should ensure the completion of the Bridge in 2 years), superstructure being supplied according to our requirements which we shall be glad to give in detail.

We have the honour to be,

Sir,

Your obedient servants,

sd/- Thomas White & Co.

Contractors.

The New Nurbudda Bridge would take four, not two, years to build. It became known as the Golden Bridge, for it was said that the cost of the build was so high that it could have been constructed in gold for the price that was paid in materials alone. Seven times the supporting pillars were washed out due to the heavy flow of the Nurbudda River.[4] But White & Co persevered, and Thomas's name is listed on a plaque secured to the Golden Bridge beneath Sir John Hawkshaw, the designer, F. Mathew, the chief engineer, and H.J.B Margrave, the resident engineer. Names written in gold on a golden bridge are the remaining trace of my ancestors in India. But the result of their time in the country lives on in grainy black-and-white photos collected by my mum, who has pored over our ancestors' history, creating expansive, confusing family trees. In these photos the men, my ancestors, pose seated in front of large colonnaded bungalows, surrounded by other men. Seated in front and standing behind, those other men are servants, more so than Thomas White & Co were ever in servitude to the British Crown and its offshoots. A hierarchy of names and bodies defines colonial India.

—

The Global South, specifically those communities and countries linked by the scars of the colonial project, is littered with stories of nabobs. There exist other bridges near Mumbai, Poona, and Surat, and on the Mula-Mutha and Tapi Rivers, that were built by men from villages nearby under the guidance of my ancestors. The family was there, in India, twice. They left the first time during the failed uprising that began in 1857, when sepoys in the East Indian Company's private army rebelled against their British commanders. The British continued to rule for nearly 100 years more, until 1947. My ancestor's second stint was 11 years long.

I do not know the names of most of the men who posed with Thomas and his sons outside their bungalows, or with the polo ponies. I do, however, know the names of the family dogs — Vic the Jack Russell, Punch the bulldog — and some of the horses: Moses was a favoured mare. Only two of the men's names are recorded on the back of a photo of my great-great-uncle astride Moses on a hunting trip somewhere near Surat: Sais and Jairam. Trusted hunting guides, valued servants. But what of the men beyond their names?

In *Creating Capabilities*, American philosopher Martha C. Nussbaum quotes the Nobel Prize–winning poet Rabindranath Tagore when she sets out to consider the role of human rights in the imperialist project.[5] Tagore defines Western culture as that which is built on a foundation of arbitrary power, without respect for humanity. In short, human rights were never ingrained in imperialist nations.

Writing of India under British rule, Nussbaum asserts:

Indians could hardly have associated empire with the idea of human rights, when what they endured every day involved mandatory segregation and denial of associated freedom;

violent, sometimes murderous, assaults on people attempting to speak and protest freely; arrest and detention without charge or trial; and other offences too numerous to list.

There's no trace of violence in my family's story; their work in India — recorded in letters and news reports — is remembered with 'deepest affection and golden opinions' by the staff of the Morvi Railway department who worked under two of the brothers, William and Charles.

But we are the creators of our own myths.

It may not be written into the journals, but the arbitrary power is evident in a farewell letter to Charles that describes him as a 'true Englishman [who] watched over our interests and sympathised with us', and in the newspaper extract that recounts William's retirement and describes him as 'a conscientious and kindly chief',[6] and in the photographs where some men are mounted on horses while others walk, and some are seated in armchairs while others squat on their haunches.

Same are named, others are not.

In the short time I lived in London, I worked for an art magazine publisher. In the understated Clerkenwell office I was reminded occasionally by colleagues that I had a peculiar outlook. My sentence structure was a little strange, my accent a little less clipped. I am from the South, a network of communities and countries deemed peripheral but bound by a history of colonialism and occupation. But I am not wholly of the South. It has always been easy for me to move up or away from the spaces, cultures, and people that bear the ecological scars of long providing for the centre of privilege, wealth, and industrial development — the North.

I got a hint of this polarity, established long ago, in 2009, soon after I returned from London, when the sky above Sydney turned red for days. The Sydney Opera House and Harbour Bridge were illuminated by a blazing backdrop that blew straight out from the desert heartland. It was the first dust storm I'd experienced. A monstrous wave of free-flowing earth that hung over the skyscrapers of Australia's largest metropolis, heralding an apocalypse few in the city could imagine, yet which was already happening four, six, eight hours inland on the barren farmland where animals were wasting away.

New South Wales, my home state, was gripped by a drought that would last until 2010 when the La Niña weather pattern would bring rain. It will be remembered as the most devastating drought since settlers started keeping records 117 years prior.[7] Since then, the dust storms have become frequent.

The lifeblood of Australia is blowing away, the septuagenarian scholar and Indigenous historian Bruce Pascoe told me over the phone one day. 'Australians should not see this as a natural phenomenon,' he says. They should be appalled. Farmers should be appalled.

Pascoe has a position at Melbourne University as Enterprise Professor in Indigenous Agriculture. But for much of the year he can be found on his property near Mallacoota in Yuin country, near the border between Victoria and New South Wales. He grows kangaroo grass there and dancing grass (mandadyan nalluk) that is let go to seed and then harvested for seed flour. Years ago, he removed the cattle that used to graze the land. He told me sustainable farming shouldn't be seen as the catchcry of greenies and radicals. It should be seen as the most conservative blue ribbon political statement on earth. 'The earth is telling us it can no longer sustain the pressure that we are putting on it,' he says. And when he says pressure, he's talking about the literal weight of tractors and grazing animals that were introduced to Australia and New Zealand by the British at the time of settlement.

Along with the animals came an imported farming system: a European system designed for the lands of Britain, with its four seasons. Down here in the South, we became nations dedicated to 'meat and three veg' built off the 'sheep's back'; a 'land of milk and honey'.

Those settler stories of breaking in the land and reaping the rewards with plates laden with lamb and beef run deep. There remains a nostalgic connection to lamb chops and beef steaks, but it is a meal as falsely endemic to Australia as Paul Hogan throwing a shrimp on the barbie. The connection remains because it's marketed as an identity artefact.

The vision of a man on a horse with a loping gait, trailing behind cattle with a dog in tow, has long buttressed the origin stories of both Australia and the United States. What is less told is that the Australian pastoralists and their stockmen and cattle were vanguard colonisers, pushing ahead of the settlers to claim grazing land from Indigenous people who fought until they could fight no more, crippled by introduced disease and weapons, and then became stockmen themselves in order to remain tied to country.[8]

Man and beast against nature is the romantic idea.

Man and beast deforesting tracts of land and pushing communities onto camps and missions so cattle could graze is the real story. Yet the myth endures, and so does the degradation. The soil is flight-ready in some parts of the country in part because of the way we've farmed this land. The simple act of ploughing loosens topsoil, readying it for the winds. Twelve-tonne tractors towing large metal discs rip seams in the soil across the expanse of the continent, preparing the land for sowing and growing Australia's big commodity crops: wheat, barley, maize, oats, canola, sunflowers, soybeans, peanuts, lupins, and chickpeas.

'Topsoil in this country is so hard to produce,' says Pascoe. 'To lose an inch of it is a crime. To lose metres, as has been the case in some places, is just pig-headedness.' In *The Biggest Estate on Earth*, historian Bill Gammage writes, 'The earth has changed. Topsoil blows away, hills slip, gullies scour, silt chokes, salt spreads, soil compacts. The last is least noticed.'[9] The last — compaction — *is* least noticed. Unless you're trying to grow food.

Sydneysiders, including me, noticed a monstrous wave of red soil because it enveloped the city for days and caused ill health among many. But few urban Australians have a physical connection to the land that grows their food. More than 85 per cent of Australians live in urban centres. But farmers notice. It's tough to farm land that's losing its fertility. And if you have a connection to that land, if you think of yourself as something close to tangata whenua, the degradation can be emotionally crippling.

When the millennium drought of the first decade of the 2000s hit, New Zealand farmers offered up holiday homes to their Australian colleagues for respite.[10] They were treating stress, but not the root cause. Farmers don't fly to New Zealand for a stress holiday anymore. Now it's different. The droughts are more frequent, the dust storms common, and many farmers are looking for change.

As the global food system developed over the twentieth century, a number of food manufacturers made a lot of money, and amassed a huge amount of trade power by making cheap food with ingredients sourced 'from nowhere' for an increasing urban population. And while various social movements may have challenged this dominance during the 1990s and 2000s — like the push against genetic modification in crops — for the most part, big food has maintained its power to define how and what we in the West eat through the adoption of green capitalism.[11]

New Zealand produces enough food to sustain 40 million people, and yet the population is only 5 million. Australia exports around 60 per cent of foodstuffs produced.[12] Simply put, those in the South produce and are aided by labour from nations further down the development ladder. Those in the North manufacture and consume. As long as agricultural wealth is centred on the maintenance of a transnational commodity trading system, the South will continue to face down the environmental impacts of a rigged deal struck long ago. Dust storms and sediment in streams are just the most obvious result.

This is the gloss of modernity, writes sociologist Hugh Campbell. Humans have the ability to physically and psychologically distance ourselves from the ecological consequences of our actions.[13] If we don't see the grey stagnant stream choked with algae, then water pollution is not real. If we don't see the cracked, barren expanse, then desertification isn't happening. If we live in a vast metropolis in the North, we don't see the South; if we live in the city, we don't see the land. But our food, your food comes from somewhere.

The smell of ripe clingstone peaches is my most enduring memory of the farm my ancestors bought in Hawkes Bay. And the feel of wheat that grew above my head and scratched at my bare legs as I followed my older sister through the golden mass. I remember ice-cold water on bare feet as I stepped across river stones, and my mum remembers her father burning the perimeter of the wheat field after harvest and letting it alone. A field put to rest.

She remembers games of tennis on the manicured grass court in front of the house, and the cool, chiselled dark wood under her hot fingertips as she traced the edge of a carved settle bench that sat in the front hall.

My grandfather would rise early when we visited to cook corn fritters, which he'd serve with poached eggs, bacon, and fried tomatoes. The fritters needed time to proof, as per the recipe that has its roots in American Indian communities from what are now the southern states of the United States. Modified and simplified by colonial cooks, it became a favourite in the family kitchen in New Zealand.

Separate eggs, and add flour, salt, and milk to the yolks. Whisked together, the batter will bind. Then add a drained can of corn kernels, and finally the stiff-peaked beaten egg whites. Fluffy and sweet, with edges that crisp in bacon fat, these fritters were my shy grandfather's expression of nurture. He'd eat half a grapefruit, cut into wedges, out of the skin and let us choose first from the feast. And then he'd be gone. Out the back door towards the farm sheds. We were afforded this life because Thomas and Mary Ann went to India.

My great-great-grandmother, Mary Ann, has been described as a 'a slim girl', 'highly cultured', and completely naive to the fact that 'life with Thomas would be a restless existence'.[14] She was taken into custody during the 1957 rebellion in India and ferried away from the 'up country station house'.[15]

She had tried to stay to be near Thomas and to keep the family intact, and I like to think she had an independent streak, a woman before her time. But in New Zealand, years later, she refused to participate when the vote was won for women. The first country in the world where women could stand at the ballot and vote, and my great-great-grandmother chose not to.

Her devotion was to family perhaps more than to progress. Three of her children died in India, two stayed on to forge their own careers and lives as engineers and civil servants; another, the eldest James, known as Jim, was left in England while the family went in search of opportunity. Six more flourished under her matriarchal wing as the family made their way south, to the farm below the Ruahine mountain range in Hawkes Bay that would in time grow to 2,300 acres, and then shrink, and then grow again, as the fortunes of generations ebbed and blossomed.

This was flax country. Vast plains once grew harakeke (flax) for harvest and use in woven basketry, matting, and ropes, and to treat wounds. The land that became the farm was once home to the Māori iwi Ngāti Kahungunu. But like so much of the land in the North Island of New Zealand, most was lost to Māori iwi through predatory land sales and confiscations by first the British Crown and then the New Zealand government.

The family purchased the land from other European landholders who 'took up' the property in the late 1860s. By the time Thomas and Mary Ann landed, New Zealand was four decades into a ruling agreement between Māori and the Crown that was outlined in nine documents that formed two versions of a treaty. One written in Māori (Titiriti o te Waitangi) the other in English (the Treaty of Waitangi), each with language that was agreeable to the signatories, neither marrying with the other. There may be reparations due to the characters who are left out of the narrative, but these details have fallen off the page or in between the translations. My family history is an intimate celebratory offering that tells the story of a family who succeeded through labour. This is colonial New Zealand.

I begin digging for worms in the paddocks behind the house in late 2018. I walk the steep face in runners and woollen socks until my shoes are caked in dust. Every 20 metres or so I stop. Placing the shovel firmly over the grassy surface, which is thatched and immovable, I jump on the edge, using my weight to sink the metal into the ground. It's hard, compacted soil.

I jump again, and again. Finally, a clod pops up with a heave of the metal face.

I pull it free from the grassy entanglement, and rip open the network of roots, looking for life. Nothing.

The silken dust falls through my fingers; despondent, I drop the clod. There's little moisture, no worms, no fungi — just matted dry roots.

I am learning that this is not how it should be. If you don't have life under the grass, how can you expect to sustain life above it?

This land I'm jumping around on is tied to a Māori hapu – a kinship group within a tribe. There are generations to come that will walk on it and expect more than what is currently here.

Pat and I came to live on this block of land with money saved from working 60- and 70-hour weeks in Sydney. That was our safety net. We did not inherit this land; it's leased from an energy company that has title over it because the land supports a vast maze of pipes, vents, and plants that collectively form a geothermal system that provides power to the national grid. And we stay here because we have the means to pay the lease fee while we learn to farm and try to turn a profit.

Below the house is a small wetland and lake fed by a creek that runs the length of the property. Black swan, mallard ducks, small black scaup ducks, and the odd paradise paddle and fight their way across the expanse. They lay eggs in the reeds and flax plants that flank the water, and lose those eggs to a passing stoat or rat, and their young to the two harrier hawks that dip and bomb from on high. The lake is nature's microcosm of colonialism: introduced and endemic species dance around each other, prodding and goading the other into action.

Occasionally, members of the local hapu will visit the farm to walk the edge of the waterway, collect watercress, or plant native tree seedlings; hebe, flax, and manuka, a tea-tree plant renowned as the source of delicious medicinal honey, are pushing up from the remains of the felled pine trees. Our time here is temporary.

This land is within Ngati Tuwharetoa's rohe, or tribal boundary. In 2017, at the end of the tribe's settlement process with the Crown, an apology was issued, ending for the tribe the process of reparations that the New Zealand government has pursued since 1975 when the Waitangi Tribunal, a body that hears grievances and advises the government, was established. It was acknowledged that the geothermal resource, the hot energy that spits from earthly vents on our farm, is a taonga, or treasure, of immeasurable spiritual and cultural importance. The apology goes on:

> The Crown regrets, profoundly, its actions, omissions and policies that have debilitated the social, cultural, spiritual, political and economic structures of Ngati Tuwharetoa. The Crown understands, and is deeply remorseful, that by removing the ability of your whanau and hapu to safeguard your whenua and taonga, your ability to nurture yourselves has been hindered.

One warm morning, we welcome a few people to our home for a hui — a conversation and sharing of information about farming and land management. One guest named Dominic sits on the outskirts of the group and listens to the conversation, barely speaking.

Weeks later, he and his cousin Wiari came back to the farm for a longer conversation. We sit at the table eating feijoa and almond cake — a simple flat cake made from ground almond meal, honey, a little flour, orange zest, separated and beaten eggs, and slices of feijoa fruit. Wiari explains that the land they farm has always been a pātaka, a source of food and sustenance, economically and metaphorically. The sustenance is in the creation of profit and opportunity for the 300 owners who trace their ancestry back to the land.

'Anything we do on the farm is about fifteen millimetres above ground level. That's where we have to make our money,' he says, as a way of trying to explain to me the difference between collective and individual responsibility. The land will never be sold, there is no inherited wealth to be had — only collective success from the profit the land produces. And that guides them.

Wiari is also a trustee for a dairy farm, and he puts me in touch with James, the farm's chairman, who tells me that they *must* balance nurturing the land with maximising profit because the benefits are vast: education scholarships, training initiatives, rehabilitation programs for troubled kids, housing opportunities, food security, the list goes on.

I had vilified the making of profit as the primary goal in farming until I spoke with James. But who am I to judge this as the main priority? Pat and I choose to value one thing over another — the environment first. But not everyone chooses the same thing. For James, the need for the opportunities that are derived from pursuing the most profit is much greater than a singular focus on the environment, for now.

At my kitchen table, Wiari and Dominic explain that the trust's productive farmland has been hard won, so it is honoured. For decades, the Crown clawed at the land, seeking to take ownership. For example, in the 1950s under the Māori Affairs Act, the Crown dictated that 'unproductive' land (defined in European farming terms) must be developed for farming or they risked losing it. When banks refused the old people development loans, because collective ownership rubbed against the need for individual debt responsibility, the families funded the breaking-in of the land by milling the tree blocks; deforestation gave them farmland and, in turn, opportunity. Milling native trees is the antithesis of today's global environmental policy, but I can't judge what I don't understand.

'It was another land grab,' Dominic says. Today they have no problem farming with a light environmental footprint because it aligns with the way their grandfathers farmed. In the summer, he tells me, the hot wind off the lake can dry the pumice soil in the western shore and create isolated drought zones. Dominic and the old people before him knew this well. They had seasonal 'country estates', as Wiari likes to joke, that would provide for them at different times of the year. Dominic says the farm follows these patterns still. Knowing the droughts will come, the stock numbers and feed budgets are modified to suit.

How long do you need to stand in place before you know when the droughts will come?

The farm in Hawkes Bay is no longer owned by my immediate family. A series of succession decisions and financial gambles means the land is now owned and farmed by cousins. But those acres carried me to where I am today: a product of privilege but not great wealth. It is privilege that gives me time to wonder about my responsibility to the land and not have to simultaneously worry about having enough to feed my family.

My mum believes privilege is never having to think about where your next meal is coming from. There was a short period when our fortunes turned, and my mum and I thought often about the scarcity of food. A sack of rejected glassy potatoes was dropped off one day by family friends. And a couple of packets of Continental pasta mix was all there was in the pantry for dinner. There were no cooked breakfasts, trays of peaches, or home baking. We relied on the kindness of others for food, for just a little while. And then my grandfather died. And inheritance followed.

Land.

Privilege.

We were safe again.

I saw the first sacred kingfisher of the year fly over the farm sheds the morning of my Big Idea. Its fluorescent-blue wing feathers and orange crest were luminous against the grey sky. Last year I spotted it a few times, popping and weaving through the tree branches along the waterways; a harbinger for what Pat calls my 'hare-brained schemes'.

After we killed the heifer for meat, I decided that we should do it again, and sell it. Cut out all the middlemen; squash the carbon footprint of transportation, processing, and export. Make beef affordable for locals and, in the process, crack the food system, just a little.

The pitch is simple, yet new: grass-fed beef processed on-farm. Made possible because the Ministry for Primary Industries (MPI) has just approved registration for New Zealand's first mobile abattoir, which means a cattle beast can be killed on the farm, processed, graded, and delivered to a butcher for breaking down. It is home-kill meat, but for sale. It is a process already in action in Australia.

To make it profitable, I ask customers to order a seasonal box of beef that includes steak, roasts, and mince, but also a range of inconvenient slow-cook cuts like brisket, oyster blade, and chuck steak. The whole animal is accounted for; the aim is no waste. And surprisingly, people buy it.

Some request add-ons: the tongue, the kidneys, stock bones. People are drawn to two things: the animal is to be killed on the farm, far removed from a high-volume, high-stress meatworks. The other is transparency. They know exactly where their meat is coming from.

But the week I'm due to deliver the first round of orders, I'm instead issuing refunds. My new business has stalled. The mobile abattoir has taken the processing truck off the road to redesign the unit to get new accreditation under MPI's Risk Management Program. I don't try another option — trucking animals to an abattoir. That would be to perpetuate the process I am trying to break. It is a failure.

Mercifully, Pat swallows his 'told you so'. The hours, days of effort to make this small venture work only to end in refunds hammers away at me. All I can see are blockages and arguments over how best to deal with farming's climate change problem. There are no shortcuts to reimagining the food system.

Maybe we should grow quinoa instead, I think. *Why not shift to growing a different form of protein?*

Pat watches me circle these schemes from afar, silently. And then he gets back to work. His tasks and changes are smaller and measurable.

One morning, I email a couple who've started to grow quinoa on the sheep and beef farm they run south of where I am. An ancient plant from the amaranth family, quinoa sustained communities in South America for thousands of years and remains a staple crop for indigenous peoples of the Andes. It's been considered a 'Neglected and Under-utilised Species' by global food policymakers despite the plant's adaptability — it grows well in agro-ecological extremes; hot, dry, frosts, high altitude, high-salt soils, you name it. And it's a source of protein, the valued commodity we trade in.

In our pumice-potted volcanic soils this ancient crop flourishes. I ask the farmers if they'll share information about the crop. They reply: 'I hate quashing your enthusiasm for a new venture on your farm …' but … they have secured exclusive variety rights from a European plant breeder to the saponin-free varieties of quinoa they grow — the variety that does well in the volcanic soils, flowers and dries off early, and requires little processing to remove the quinoa's typical bitterness.

They have invested in the crop and produced and packaged small bags for home cooks under their own brand name. They've assumed all the risk for developing a new product line and finding a customer base for it, and for now they're doing it alone. They suggest garlic or blueberries instead; both do well in this growing zone. So, I start researching garlic varieties, and look for seed banks that sell rocambole and soft-neck varieties. I search, only to find waitlists, which I duly join.

New Zealand grows garlic, but only just. There are only four commercial garlic farms operating; the supermarkets are awash with imported Chinese garlic that sells for four times less than the locally grown bulbs. I miss the planting season.

Instead, I start throwing a seed mix over a paddock chewed down by our cattle just before the last rain. The money that might have gone to our first small trial crop of quinoa or garlic is instead spent on 60 kilograms of diverse pasture seeds: fescue grass, cocksfoot, chicory, plantain, clovers, lucerne, vetch, lupins, rape, radish, mustard, linseed, and phacelia. I read that this cocktail will grow herbage as feed for our cattle, and provide nectar for the bees that bounce from phacelia head to red clover flower and on to the kowhai.

'Just stop,' I hear Pat say from somewhere far ahead of me. But I don't listen to him. I am determined that I need to change everything now, and quickly, to prove that it's possible. I desperately want to be the hero of our story; to save something, someone. I didn't understand then what Dominic had been trying to share as he sat quietly on the edge of the group. The decisions we make now are for the benefit of our grandchildren, the generations to come. There's no glory, no reward to be collected now.

It'll take 20 years, Hamish Bielski had told me, for people to change.

It's not fast, I thought.

Hurling the seeds by hand, arm stretched long, they bounce and float by the hundreds atop the grasses already present in the paddock. It didn't occur to me that contact with soil was needed. I throw and throw. Each seed a new idea designed to fix something.

Each seed a false fix.

5.

Meat money

There's a pen mark on the otherwise clean whiteboard propped against the back wall behind where my accountant sits at his company's board table. It's distracting me from what he's saying: 'Technically, you're poor'. The money we earned in Sydney now exists as cattle or is gone, wasted on expensive milk powder. There is no cash flow, and our savings have diminished. And we are heading into the cold months when we can't or don't trade cattle; no one is buying at this time of the year, especially not so soon after the summer drought.

'I should never have come,' Pat mumbles when we're alone in the truck outside the accountant's office. It's soft — he's never quite able to say it clearly or commit to the comment, the implied leaving of it all.

This version of farming life, with its rolling disappointments and drudgery, wasn't what we'd envisioned. Not for the first time we look at each other and ask: *What* should *we be doing on this land?* The answer is *less.* We have no other option; our money is all but gone.

—

Soon, we start hearing news of a contagious virus spreading in the Chinese province of Wuhan. For weeks we've been worried about the bushfires that blasted through Australia in the hottest and driest year on record, destroying more than 24 million hectares and displacing millions of animals who fled ahead of fire fronts so furious and vast that the smoke clouds stretched thousands of kilometres.

Pat's parents sheltered inside their Sydney apartment, away from the choking smoke, for most of the summer. A Royal Commission would eventually report that natural disasters are likely to become so complex in the near future, as the earth continues to warm, that traditional firefighting techniques just won't cut it.[1] This complexity is the result of climate change, whether politicians admit it or not.

So Sydney's residents are already walking the streets wearing face masks when, on 31 December 2019, the World Health Organization (WHO) is alerted to the presence of a rapidly spreading airborne virus. The virus jumps to more than 200 territories around the world. Countries shut their borders, governments lurch between temporary shelter-in-place measures to strict lockdown regulations that prevent citizen movement, all to curb the spread and slow the deaths. New Zealand's government introduces some of the strictest measures in the world, and at 11:59pm on Wednesday 25 March 2020, the entire country is asked to stay home.

We keep working; farmers are designated essential workers. Food producers, regardless of whether they are growing food for export or local consumption, are required to keep going. But the strict measures of the lockdown are barely felt here on the farm. We rarely leave anyway. We work seven days a week, our ability to travel and enjoy life beyond the farm gate tempered by our vocation and our financial limitations. The glow of the early promises of easy profit is long gone. Six months is all we have left in us. If we don't work out how to make money from this land, how to survive as farmers before then, we'll be done.

Entirely.

We are at the whim of the market, climate change, and another virus that was present on the farm and in our animals long before COVID-19 descended on the country and the world.

In the deep July chill, I awaken and draw my baby daughter close. Soon we're both dressed in layers of wool and waterproof clothing, and we walk out the back gate, under the pines, and towards the shed; she in a buggy sheltered by a plastic rain cover, me pushing and bracing for the hours ahead.

This routine is well practiced; I'm rearing calves for the fourth season in a row. Pat has risen hours earlier, and is already out in the paddock, standing spread-legged, arms folded, defiant or distressed depending on whether the calves are suckling. A plastic feeder system on wheels, with 50 black rubber teats, each shaped in the form of a cow's teat, and a 300-litre tank are hitched to the quad bike.

Calves are squeezed together around the feeder, suckling the warm milk mix as fast as they can, their hindquarters quivering as the warmth filters through their small bodies. Four mobs of calves; four feeding stops.

The calves I'm in charge of are the small ones, the sickly, the very young. And they are housed in the refurbished hay shed, which has been extended with long bark chip–floored runs that poke out into the yard, giving the calves space to gallop and twirl the length of the pen. But most don't gallop or twirl. No level of cleaning can rid us of the bug that's in the shed and takes a small animal down within 24 hours.

We have a new vet, an American woman around the same age as me, who works mostly on the dairy farms in the region. One morning she drives into the compound with an assistant to attend to seven of our calves. Two white-head bull calves; a large fluffy chocolate-coloured Angus-cross that I've named Bear; a Friesian-Hereford heifer; and three small jet-black, rail-thin Angus-cross bulls.

All are lethargic, barely eating or suckling, and they're rapidly losing weight; their flanks cave in, leaving their hip bones protruding like hand grips. They'd all been in the paddock for weeks, starting to nibble on grass, racing the other calves the length of the paddock in the late-afternoon witching hour, when calves feel a collective urge to play.

The vet takes their temperatures; all are elevated. She draws vials of blood. Before she leaves, each calf is given a large shot of broad-spectrum antibiotic, and I record the details, matching the calves' ear-tag identity numbers with their treatment details. I choose to use antibiotics, but I don't have to. The use of antibiotics on livestock is contentious. More than 90 per cent of cattle in Australian feedlots are given antibiotics at some point, often just once, as a preventative measure against diseases such as bovine respiratory disease. But it adds up.

In *On Eating Meat*, Australian food writer Matthew Evans reflects on the work of biologist Rob Wallace and writes: 'The more antibiotics we use, the more resistant other bugs will eventually (and irreversibly) become against those antibiotics and the more likely it is that we'll have a catastrophic outbreak of disease. The worst-case scenario is a superbug that may cross species not just into other livestock, but — terrifyingly — humans.'[2]

The worst-case scenario has arrived. We don't know what's killing them, but we want them to survive. I judge myself, but the condemnation from the vet is worse.

'She really shouldn't be down here,' the vet says, looking at my daughter's face pressed up against the plastic buggy covering. My daughter is curious to see what's happening beyond her plastic-covered enclosure. 'I had a girlfriend who had a car seat in her work truck and her two-year-old picked up rotavirus.' It's so easy for the virus to jump, the vet says. 'If you touch the buggy with gloves on you might leave a trace.' And if she sucks on the bar, like she does, then that might be enough.

Contagion.

Transmission.

Sickness.

But we don't yet know what's making the calves unwell. That's what the blood tests are for. It shouldn't matter, says the vet. 'She shouldn't be here.'

I can't do it all. I can't be in the shed with the sick calves, keep my daughter well, and keep the animals well. Maintain a system that's intended to make us money. Imagine a new version of the farm that will carry us into the future. And be part of a 'we' that has disappeared in the fog of seven-days-a-week-work and struggle.

The blood tests come back inconclusive. The vet doesn't know. It might be *E. coli*, but the symptoms aren't right. It's not rotavirus. It's not salmonella, but it could be. Or *campylobacter*.

The stream that flanks the farm and runs into the small lake below the house is crowded now with flourishing flax plants, native grasses, nikau palms, ferns, and hebe shrubs. Purposefully poisoned to make way for the natives, the willows have become brittle and grey. The wetlands are flush with birds. And the pukeko — native swamp hens — have bred and become confident, stretching their gangly red legs and stunted midnight-blue wings to pop and duck over the fences and under the wires, stealing morsels from the calf-feed troughs that sit on the rails of the shed. In farming lore, these resplendent birds are considered agricultural pests for carrying salmonella, destroying crops, and fouling pastures and water troughs. So, they were shot by farmers. In large numbers.[3]

But the source of the animal sickness might also be the three white pekin, one mallard, and one muscovy duck that have been introduced into the wetland system and have become farm residents, waddling bolt upright from the lake, up the hill under the steel gate, onto the driveway, and through the six-foot wire deer fence that separates the calf paddocks from the house.[4] There they graze, eating grass and calf meal. Growing fat. Content to be part of the domesticated menagerie.

I'll never know for sure if the birds made our calves sick by contaminating the feed because the vet's words dropped cold in front of me: 'She shouldn't be here'.

I caved inside. And I stopped. Finally hearing Pat's words.

Bear the calf lived. So, too, did one of the black Angus bulls, and the heifer and two bulls. The others died.

Our daughter remained safe.

As COVID-19 descended, our life on the farm continued, because the tremor caused by a system lurching into flux from the fallout of another unknown disease had already rippled through. Sickness had forced us to confront a farm out of step with nature. We need to learn to farm *with* the pukeko, not against. To farm without adding a synthetic layer to this land because our reserves are spent. But how? How? When we have nothing left to give?

Pat takes a call one morning, early, around 7.20am. It's an opportunity to capitalise on last year's work, an offer made at a time when we most need money. And so Pat listens.

Are we interested in selling our yearling heifers?

For live export?

Rarely a month goes by when we don't pause to search our moral guidebook, but scripted answers are infrequent.

When this offer is made, New Zealand has temporarily halted live exports of cattle. On 2 September 2020, a vessel, the *Gulf Livestock 1*, which is owned by UAE-based company Gulf

Navigation Holdings Ltd and sails under the Panama flag, sank off the coast of Japan in typhoon weather. On board was a crew of 43 sailors, vets, and livestock handlers, and almost 6,000 New Zealand dairy and beef heifers, much like ours. Three men were found, and two survived — one found in the water, the other on a life raft. The rest were gone. Yet a search continued, funded by distraught family members. There was no search for live cattle; all drowned or died as the ship rolled and broke violently in the swirling weather.

Few realised at the time of the sinking that New Zealand operates a live export trade for cattle, the sheep equivalent having been long ago banned due to animal welfare concerns. Yet New Zealand ships almost NZ$500 million worth of cattle each year, all breeding stock heading mostly to China. They don't go to abattoirs abroad — not immediately, anyway. They go, we're told, to farms that are 'very, very well run'. And the price we are offered is vastly higher than anything we can get on the local market: 70c per kilogram more.

We must decide if we can shoulder the moral burden of shipping cattle over a vast ocean for 17 days to China, so that the country's farming corporations can breed beef herds to keep up with the rising demand for red meat from the flourishing middle-class economy. The juggernaut of emotional and philosophical dilemmas I face is stretching my resilience. If we say yes, is the journey the cattle will take worse than death?

Spring has arrived, and the storms roll over the lake and onto the farm. The soil, lacking moisture, sucks up the rain. The country is dry, but the grass is starting to grow; so too, miraculously, are the seeds I threw out feverishly over the small paddock: vetch, phacelia, radish, mustard, lucerne, fescue grasses, clover. We spread the last round of synthetic fertiliser nine months ago, and in the shed awaits two large tubs of seaweed fertiliser made from bull kelp collected from the coastlines around New Zealand. Another practical step that takes us away from conventional farming and extravagant expenditure towards something different. But each day we must push to stay afloat financially.

One spring night, snow starts falling on the farms south of us. A blast is chilling the country, unusual for the time of year. On the computer screen, a South African farmer turned tutor taps and huffs, trying to share a PowerPoint presentation with the 48 farmers logged into the Zoom call.

It's the era of the Zoom Boom: the knowledge economy has left the stage and taken root on a screen, in a video communications company that over the course of the COVID-19 pandemic becomes a $40 billion company worth more than seven of the world's largest airline companies combined.[5] And a 70-year-old cattle farmer is learning how to use it in front of us. He is a farming coach, and his wisdom is simple: 'Just start', 'diversity is stability', but also, 'if you're broke, you can't save the soil'.

He's a practical not ideological guide who understands our need to make money. So, we listen, surrounded by a virtual community of farmers chorusing the line that we are here to do good; to give back; to nurture, not degrade, by farming cattle. But when our cattle and product leave the farm, I'm sucked back into the polarisation that's still driving our food system, and it's getting worse.

New Zealand and Australia's traditional trading partners are in a state of high turmoil. The United States' beef industry is awash with COVID-19, and abattoirs are shuttered as the food system is exposed as a space of worker exploitation. A reckoning is occurring. Trade with China has become a fraught game of ping-pong as human rights bounce off export dollars; and Britain is in the throes of a local-farmers-first campaign as free trade deals are brokered to ensure a steady supply of food into a system that relies on imports, the UK government having long ago bet on cheap food. COVID-19 and trade deals have captured the attention of the masses. And

I think of what Pat's niece told me: 'Climate change was more important last summer.'

Except to those working on the land. Here it is real and urgent every day.

In *The Fate of Food*, journalist Amanda Little considers the way food farming is changing in this era of climate strife. She meets a vertical farmer named Ed Harwood who tells her that the world will need a lot more farmers to feed the growing global population. But it's a different type of farmer. 'Growing sustainable food will increasingly require dynamic teams of people with the kind of technical skill sets that young urbanites may be more likely to have than seasoned heartland farmers,' she writes.[6]

Who among the young urbanites is going to lead the change we need? Food tech entrepreneurs? Cooperative organisers? New food retailers? Or those who venture onto or back to the land?

If you head inland from Sydney for long enough, over the Blue Mountains, through the university town of Bathurst and west towards Dubbo, the horizon will eventually flatten and expose the vastness of Australia.

I have no connection to this space. If anything, it exposes an underlying sense of anxiety because I can't read it. English naturalist and journalist Michael McCarthy explains that most humans are hardwired to seek the edges and cover because nature is 'not paradise'.[7] Yet nature is also where we source some of our most powerful metaphors, so the fear is just unknowing.

Murray Hemi, who held the role of general manager for environmental leadership for the Māori milk company Miraka when we first met, told me that when people stay in a place long enough, the land will start to permeate their being. And in turn that person can read the land, stand for it, and be its voice. I'm driving towards the town of Eugowra in New South Wales to meet someone who claims to speak for a small part of this place.

The dry lingered over summer and into winter here in New South Wales, as it did in parts of New Zealand. The adage 'it always rains' is on high repeat back home, because we waited far longer than we usually would for the storm cycle that comes to break the summer heat. The farm is still losing money. The offer to sell our cattle via live export clouds our conversations. We can start selling off young cattle for little more than we purchased them for as tiny calves; that's an option, too. We're at the whim of the market and the weather — which is farming, I've been told.

I've flown to Australia to better understand a type of farming that's thought of as drought-proof, intensive, and very much like the kind taking place at the farms our heifers are heading to if we say yes to live export. My baby daughter is asleep in the back of the rental car; she will stay that way for two, three hours if I'm lucky. Beside me is a friend: moral support and babysitter.

The distance on the map between Sydney and Eugowra is the size of my thumb; it seems short. On the map. But nowhere is close once you get out of the city into flat country. I drive, for hours and hours. And my daughter starts to cry, then scream. I know why I'm here. To learn. To figure out how to farm more efficiently, effectively, profitably in a hotter climate. Pat doesn't understand why I've come. What answers lie elsewhere? My trips, my conversations are me on my knees, scraping back the dirt, looking for something lost. I may never find it, but I believe the act of looking is worth the effort.

Finally, we arrive.

There's a cafe in Eugowra that takes pride in being in cattle country. Cow figurines line the shelves inside, while outside a large novelty cow crowns the front door. A grey-haired, jovial man stares quizzically as I order black coffee. His side eye reveals his intent. He's trying to place me. He's reading my face, looking for a trace of someone from these parts. Why would someone be in Eugowra if not to visit a family member?

Since I was a toddler, I've been told I look like my father: the same complexion, eye shape, nose. I didn't miss this familial connection when I moved abroad, but I understand the need some have to place a person. In small towns and rural parts of the country, outsiders might misjudge, misinterpret, misunderstand something. And the urban–rural divide continues. The cafe owner gives up and asks: 'Where are you from?'

New Zealand, I say.

And off he goes. 'I once knew a couple of fellas from New Zealand. Farmers. Good farmers. Cattle men.' And on it goes. I drink my coffee. It's burnt but hot and will suffice after the long drive.

Outside, the galahs are calling from the powerlines that stretch above the road. Two more men are sitting on a roadside bench over the road. They, too, are staring. Not a lot of visitors pass through here, I guess.

About 20 minutes back down the road, I'd passed a farmer on a two-wheel motorbike slowly trailing a couple of Hereford cows and calves as they grazed the wiry fodder along the roadside.

Behind the fence was barren dirt; a cluster of grass stubble and the trace of a ring of hay were visible. This is what cattle farming in drought is: chasing every blade of grass and fodder down the highway and across the range. The cows were thinner than they probably should be with calves in tow, but they are fed. Just.

Thirty minutes beyond Eugowra, I find fat cattle.

My arrival at Gundamain Feedlot is announced by a bell over the door and the echo of my boots on wooden floorboards. Two women sit with their backs to me, working on computers, looking out over the flat land beyond. A handful of cattle laze in a paddock behind the office; some sniff at a pile of hay.

From the side office a blond woman, hair in a bob, steps out into the room, turns, and shakes the hand of the man trailing behind her. She acknowledges me and I wait. This woman is in charge, and so she must be Tess Herbert, who I'm here to meet. Tess and her husband own and manage Gundamain Feedlot, a 6,000-head farm that's been in the family for 150 years. I'm told it's a typical operation, but I suspect it's not. For starters, it's run by a woman.

There are around 450 registered feedlots in Australia; more exist and operate beyond the realm of regulation. Farms that house less than a thousand cattle aren't required to register with the Feedlot Association.

A meat lobby group, Future Beef, estimates that 80 per cent of beef sold in Australia's major domestic supermarkets come through the feedlot system.[8] And more is exported as prime beef to parts of the world that value 'marbling', the term used to describe cuts of beef that are shot through with fat veins. It makes the meat tender, apparently, and is the result of the high-energy diet cattle are fed while in the lots.

Yet the vision of a feedlot sits uneasily with a lot of consumers: thousands of animals penned in areas little larger than a football field. Here, it's 100 cattle per pen. Fifteen square metres per head. They eat and chew their cud. Then gravitate to the shade cloth that runs as a strip through the pens. That's life.

Not all feedlots have shade, despite temperatures rising to 40 or 45 degrees Celsius out here over summer. It's an added cost that some operators don't see fit to shoulder. But the good feedlot-ers, like Tess, consider it an animal welfare issue. 'Cattle can cope with heat,' she says. It's the stillness plus the heat that makes them distressed. So, the installation of shade cloth helps her sleep better at night.

I wonder how many other feedlot owners have trouble sleeping?

Tess is an amicable host. As we exit the office, we're accompanied by a fox terrier and legions of flies. It's rained a little recently, and the flies have proliferated. They hover near my nose, and I swat as we walk up the dirt road towards a series of large silos, the wind whipping up the disturbed dust in our wake.

She chats about her office staff: mostly women she's managed to keep because Gundamain is a family-friendly business — babies are born, childcare is provided, maternity leave provided, and the women stay working in an industry dominated by men. I enjoy her company but deplore her business. In those first few minutes on the feedlot, I can't see past the intensity: so many cattle in such a small area, so much feed trucked in, so much activity. I'm repulsed and yet, what Pat and I were doing rearing calves in pens in a shed is not so dissimilar. I swallow my judgement for fear of revealing myself.

'It's like a motel for cattle,' Tess tells me. And it only works, financially, if the motel is full. So, every week cattle are trucked in and out. In from grazing farms in the region, out to meat-processing plants.

The cattle are here at Gundamain for 70–100 days, but they spend the first year of their life grazing elsewhere. Tess tells me they work mostly with graziers in the region, and that means these cattle are coming from drought-stricken farms. The notion that they're grazing is tempered by the reality that there's near to no grass growing anywhere near here.

When they arrive, Tess and her team manage the transition into the penned lots slowly. First, the cattle are put into paddocks and the feeding regime begins: small doses of the high-energy grains they'll come to consume in large quantities are fed with hay. They get used to the other cattle around them. It's a form of acculturation. Dominance is asserted, rules are obeyed. Men on horses ride through the mobs, looking for displays of aggression and spotting those who grow weak or sick in the new environment. Soon they'll be moved into the main pens, and they'll continue the routine, eating a mix of cotton seed, canola seed, hay, silage, cereal grains, and barley. This is the opposite of a free-range, grass-fed farming system — the romantic vision of farming. But that's not to say it's worse. In this part of the world at least.

Marco Springmann, the Oxford University public health researcher who champions the term flexitarian, is adamant that the meat industry's focus on grass-fed beef is obfuscating the true impacts of the beef sector on the climate. 'Lots of people have this romantic attachment to pastures and grass, but that really should not be so,' he says.[9]

The notion that grass-fed cattle is better persists, but so, too, does the chorus of voices that insists grass is not always better for the cattle or the environment. Isabella Tree, a British farmer, conservationist, and author of the book *Wilding*, suggests that cattle flourish best eating a range of grass and roughage. Put them onto a sweet, energy-rich ryegrass pasture and they'll gorge themselves and suffer for it, with bloat and constant belching to follow. I've heard this from an Australian dairy farmer, too, whose cows baulked at a flush of deep-green grass made heavy with nitrogen fertiliser.[10]

Ryegrass is the dominant introduced grass species on New Zealand farms. Millions of dairy cows and beef cattle are sustained on a diet of mostly ryegrass, and we claim a superior form of farming. But is that diet contributing more to the methane count? Marco Springmann and his Oxford University colleagues say yes.[11] A cow, he says, that is freely roaming on sown pasture emits roughly double the methane emissions than a cow in an intensive meat-dairy system, a system like Tess's. 'Sometimes it can be really unintuitive where the impacts are,' he says.

In 2017, Australia's peak scientific body, CSIRO, collaborated with Meat and Livestock Australia to devise a pathway for the industry to hit carbon neutrality by 2030. The feedlot sector was a crucial part of the plan because the system can get cattle to the kill weights much quicker and more efficiently, so the animal spends less time on earth, farting and burping. And the manure waste produced can be stored in enormous effluent ponds where methane recovery technology can transform the methane gas emitted into electricity, or iron sulphate can be added to the ponds to drop the greenhouse gas emissions yet again.

If you're looking at just greenhouse gas emissions, sending cattle to feedlots is better for the environment.[12] Better yet are intensive pig and poultry farms, according to Springmann and other researchers at the Oxford Martin School, who conclude that one kilogram of beef produced in a 'normal' beef unit emits about ten times more greenhouse gas emissions than a kilogram of either pork or chicken.

But environmental gains are not just about numbers.

In the United Kingdom, intensive pork and poultry operations are on the rise to keep up with demand for the 20 million birds slaughtered each week. The number of intensive farms is almost 1,800 and climbing, up from 1,669 in 2017.[13] Seven out of ten of Britain's largest farms house more than a million birds. It's a trend being replicated across the West. In New Zealand, the 'intolerable smell' was a binding enough insult to bring 'retired folk, life-style block owners, the elderly, and single parents' out onto the streets to protest[14] when in 2018 chicken producer Tegals proposed a 23-shed chicken farm housing 1.3 million birds in a part of the country desperate for jobs.

In eastern Oklahoma — a part of the United States rife with chicken farms — writer Rebecca Nagel, a citizen of the Cherokee Nation, says 'you can smell the chicken poop, the ammonia just hangs in the air'. She talks of algae problems in the streams and rivers, and of watching the sky slowly change colour over the years as the chicken houses proliferate.[15]

Yet it's this scene that food policymakers concluded would

lessen the environmental impact of keeping the world in meat. Beef emits ten times more greenhouse gases than lower-impact meat, like chicken, which yet still emits about ten times more greenhouse gases than plant-based protein sources like legumes — chickpeas, peas, and beans. It's all just a numbers game — footprints, profit margins, feed budgets. So how do we account for ethical or cultural impacts of intensive meat farming? Are they always secondary to environmental gains?

Walking along the feed pens, following Tess and her little dog, the idea that Pat and I should also aim to kill our animals before they hit two years of age strikes a chord. Heavy cattle can send topsoil and sediment rolling down our steep paddocks and into the waterway below just because of their weight. Heavy cattle demand more winter feed, more hay, silage, more winter cropping. Heavy cattle will continue to emit methane, and that is our focus. So, what if we had cattle that never became heavy cattle? Can I sleep well at night if I kill our animals for meat before they've lived a full life?

Not long after I visited Gundamain, I spoke with Professor Clive Phillips, an academic from Queensland University who chairs the Queensland Government's Animal Welfare Advisory Board. And I wondered if the answer was to stop farming cattle altogether in this part of the country or to bundle them into feedlots to be as efficient as possible. Professor Phillips has travelled extensively throughout Australia, visiting farms and corporate-run cattle operations to gauge whether animal welfare standards are being observed. And he's reached the conclusion that we — Australia, the world — don't *need* a cattle industry.

An argument I've heard often is that the essential element of farming is not likely found at the point of consumption. We don't need beef for our nutritional wellbeing; that was made clear in my conversations with Dr Anne Louise Heath. 'There is no such thing as vitamin beef,' she told me.

But red meat is valued beyond the plate for its economic worth. Is it valuable enough as a jobs creator to offset the environmental impacts or concerns voiced by Phillips about collective animal welfare standards?

If you consider yourself an ethical consumer, at some point you've weighed a set of values and chosen one over the other. Not everyone is going to try to mitigate climate change while eating dinner, but they might be trying to support a foodscape that provides a living wage for individuals.

Eugowra, the town where Gundamain is based, has a population of 779, according to the most recent Australia census data from 2016. While I was at the feedlot I met or saw six workers, many of whom were women. And they are working mothers. In a town that size, there are few options for employment. Is that the value that makes feedlot farming in this small town essential or is there an alternative?

Further out past Tess Herbert's farm live Karin Stark and her partner, Jon Elder, a fourth-generation farmer whose family have been on the land for a hundred years. Their 2,500-hectare cropping property is west of Dubbo in New South Wales, and they haven't harvested a winter crop in two years. The cotton has always grown in summer with the aid of irrigation, and diesel-powered pumps that push the groundwater to the surface, letting it flow out into a reservoir and then into the laser-flattened cropping paddocks via rows of precise farrows. But the winter chickpeas planted in dry land wouldn't take. The reason is simple: 'There's nothing falling from the sky,' says John. 'We've just experienced not the longest drought on record, but the sharpest one, and it was off the scale.'

Karin says in recent years the dust storms have come every few months, and for a while, every few weeks. It took moving to the farm in 2011 for the Perth-raised environmental consultant to truly comprehend the impact of climate change on the Australian landscape. 'Our topsoil was being blown away,' she says.

I am the same. I didn't understand how the land was changing until I had dirt under my fingertips. I've often thought recently that if the surface of the land were a vast sheet of tracing paper, perhaps our collective outrage over food and farming would dissipate in a sense of awe. And from that gasp of wonder might come hope, not argument.

But it's not transparent. Arguments and protests continue, and hope is hard to find. But it is there.

Karin and John consider themselves conservation farmers; like 5,000 others around Australia, they're part of the lobbying group Farmers for Climate Action. For years they've been pleading for the Australian government to prioritise climate change in agricultural policy. John is fearful that the world is moving too slowly. 'I am just befuddled by the whole attitude,' he says, elaborating that the denials and defensive position from Australia's governing politicians are despair-inducing. They expect support from rural communities come election time, yet it is farmers who are on the frontline of climate change.

And so they make their own way forward. Not waiting for regulations.

To prepare for a future in farming they're experimenting with renewable energy in a part of the country rich with sunshine. They switched the diesel pump to solar, and in the first year they saved around A$180,000 in fuel costs. They were the first cotton farm in Australia to be accredited as a power station because of their solar system and were issued certificates for the reserve energy produced. These intangible treasures have a value on the market and can be sold to heavy-emitting companies to offset their footprint.

They have ambitious plans to harvest sunshine as an additional income source, but being an early adopter is risky. Karin tells me they've had technical difficulties blending the diesel pump with the solar panels to create a safety net during cloudy periods. They need technology to catch up with their vision: a vanguard problem not limited to solar farming. But they remain undeterred, and when I ask John how he feels about becoming a sun farmer, he laughs a little before saying: 'Well I guess I'd say I'm indifferent.' Not that he doesn't care — just that there's no unwavering, blinding loyalty to cotton, wheat, or beef above all else. His identity is not wedded to farming just one thing.

As I sat across from Tess Herbert in her office, the air conditioning unit humming behind us, I asked if she thinks we need to drastically change the way people farm on this land. She said yes. But with a caveat. 'Large parts are ideally suited to livestock grazing, so that's the rangeland areas of Australia where you can't grow plants.' This is the same rangeland historian and author Bruce Pascoe says is ideal for free-range farming; of kangaroos, soft-footed animals that produce good meat. It's the same land where John and Karin grow solar panels. And it's the rangeland that Professor Michael Carolan pictures when he thinks about a new form of farming that stands in opposition to corporate power structures and private ownership: he sees a future of collectivisation, a farming system performed with others in response to the specifics of the community and the land. In economic terms, if you share land, machinery, and resources, your costs go down. But culturally and legally, this is a complete change for farmers with Anglo roots. Succession is broken, individualism is sidelined for cooperation, and corporate structures are skewed.

The largest landholders in Australia are corporate farming operations, some overseen by financiers and billionaires with new ties to pastoralism, like mining giant Gina Rinehart, who owns 9.7 million hectares of grazing land; others trace their roots back to the late 1800s and run deep blue in political persuasion, like South Australia's MacLachlan clan, who own Jumbuck Pastoral Company with its 4.9 million hectares.[16]

Rip down the fences, says Pascoe, we'll be better off for it. Except those who own all the land.

Back in New Zealand, farmers have started protesting about everything, which makes the gesture count for nothing. Publicly they're calling to halt freshwater regulations and pump the brakes on climate-change mitigation. But what the protest is really about is tradition, and property rights — the perception that the government is trying to encroach on a farmer's land, land often inherited along with its value and wealth, generation after generation. It's hard to be sympathetic to the cause when the target of the protest is environmental protections.

More than 40,000 people follow the protest group Groundswell on Meta's Facebook. The group was started by Bryce McKenzie and Laurie Paterson, a fifth-generation septuagenarian beef and sheep farmer from the South Island. Many more turned out when the group staged nationwide protests railing against the government's freshwater regulations, electric vehicle initiatives, and so-called 'land grabs' — a reference to local councils attempting to map significant natural areas on private land for biodiversity protection. Groundswell advises landowners to do all they can to keep the mappers out. Lock the gates, don't let them near 'your land'. There have been calls for the founders to start a political party.

They aren't anti-environmental regulation, the founders claim. They just don't like 'unworkable' regulations, preferring instead localised, collective action to improve fresh water and biodiversity rather than one-size-fits-all national regulations. Many New Zealand farmers don't consider this a contradictory position.

Pat and I attended the first protest, out of curiosity rather than support. I asked one woman sitting in a black Dodge Ram truck why she was joining the 200 or so protestors. She pulled out her phone and found an apocryphal quote often attributed to Thomas

Jefferson, a founding father and early president of the United States: 'When the people fear the government, that's tyranny; when the government fears the people, that's freedom.' Her son Hori sat in the driver's seat. He has land nearby and interests in kiwifruit and forestry, and he was there to protest the 'land grabs'.

Another couple, Jac and Simon, told me the regulations felt like 'death by a thousand cuts'. The value of their land has dropped because it sits within a catchment area that limits land and nitrogen use under the Regional Council's cap and trade program. The only thing that will return their asset to its former value is to break it up into parcels for lifestyle blocks — that is, to quit farming. Jac is resolute that's the last resort. 'We are not trying to stop environmental improvements,' she told me. 'But they need to be equitable.'

We were interrupted by an elderly man handing out bumper stickers from a little bum bag. 'Will you put this on your truck?' he asked Jac and Simon. The sticker read 'Jacinda, Get Out of My Life'. Simon accepted it and rolled it uncertainly in his palm.

All this because in September 2020, the New Zealand's National Environmental Standards for Freshwater came into force, aiming to halt lasting damage to the country's 425,000 kilometres of riverways and 5,000-odd lakes. It requires farmers to plant wide tracks of flax and other native species around wetlands and streams, and fence off waterways from cattle, sheep, and deer. Additional measures have been mooted and rolled back, often following complaints from farmers that the rules are unworkable in practice.

These regulations reflect what I believe in: a future for farming that's focused on improving, not degrading the environment. A future for all. But I believe in the concept of collectivism — we're all in it together — which is the opposite of inherited family wealth and individual title.

It's always about money, even when it's not.

In 2021, the median price per hectare for grazing land like ours was NZ$11,115. If we set out to buy the lease block we farm, it would cost us around $1.5 million, but that's only if it was kept in use as a grazing farm. Around us are dairy farms, and the prospect of this block becoming yet another dairy farm conversion is great. So, too, is the increase in value. It triples the price of the land. Suddenly the asking price might be closer to $4.6 million.

Pat and I spoke about trying to buy land once. It was a conversation that lasted for a minute. No amount of passion and will can gather the millions of dollars needed.

As my dad's generation protest, farmers and food producers like us are being shut out of land ownership because it is, as it has long been, the perfect representation of intergenerational wealth and power. Those who have it don't want to give that up. Those without, like us, are trying to break the hierarchies.

In Australia, corporations dominate the big end of farming and food production, and that means the number of stewards on the land — people who have a vested interest in preserving it for future generations — is dropping. The land and climate are changing, the regulatory environment is changing, consumers' demands are changing, and farming is too hard. But what if farming became something other than what it is today?

Sam McIvor thinks that's possible.

He's a farm kid without a farm. And he's seen the thriving country communities of his childhood drain through corporatisation and urban drift. He is adamant that the counter is to put the next generation of Australian farmers back on the land.

Sam's come up with a social enterprise system that pairs the next generation with retiring farmers in a profit-share partnership, and woos urban investors keen to have a direct connection to their food and its farmers. His organisation, Cultivate Farms, is, he says, ultimately about reviving ailing communities, but it's also a gesture of collectivism, whether Sam intended it or not.

When you start thinking about farming in a new way, and it involves collaborative and shared ownership, then it also means the sharing of expenses. New Zealand has a long history of cooperation inspired by the Rochdale Pioneers — a group of nineteenth-century mill workers who pooled resources for their community to purchase basic foodstuffs at an affordable rate. Shoppers became members, and the co-op was born. Some of New Zealand's largest food and agriculture companies — Fonterra, Food Works, Alliance, Ravensdown, Ballance — started as farmer cooperatives. But the 'all in it together' mentality has long been replaced by the pursuit of global trade and shareholder profit.

Yet collectivism can change the bottom line and allow people to think about what can and can't be done based on what's good for the environment as well as what's profitable.

What Pat and I do, and what we see in rural areas around Australia and New Zealand, is a European style of farming that requires fences, domesticated livestock, and land managers. What would Australian or New Zealand farms look like if designed specifically for the land we have? What food would they grow?

Tess Herbet, the feedlot owner, is inspired by calls to reimagine what the agricultural landscape can look like in the years to come. Bruce Pascoe says, 'It's going to be a different style of farming,' and Tess Herbert responds: 'It isn't necessarily a cause of fear.'

I head home. Away from the red centre and back to the farm. On the flight back to Auckland, my baby girl snuggles into the crook of my arm, feeds, and then falls asleep. I drop my head back against the headrest and listen to the man next to me talk about the sunsets over the water off the coast of Darwin. He's a boat captain who works in the gas mining industry. For months he lives on a ship, circling a liquid petroleum gas well in a pristine part of the world.

And then he leaves and goes home to small-town New Zealand for a rest. Humans have the capacity to remove themselves from the site of their industry. It's a mark of modernism. I, too, can travel across the Tasman talking to others about the land they manage and farm, but nothing will change for us until I stop moving.

The Australian artist Janet Laurence once said she came to understand the land — the subject of so much of her work — by looking 'through it'; looking inside it.[17] And the result has always been chillingly beautiful. She was once part of a program funded by the Queensland Government called the flying art school. With fellow artist, potter Gwyn Hanssen Pigott, Laurence would fly across remote Queensland and the Northern Territory, visiting with students learning to paint and pot by correspondence. And on those trips, she would cast her gaze downwards to the land — red in parts and crisscrossed with the tracks of animals and trucks, lush green in others, where the stems of a waterway pushed moisture over onto the land — and she realised her students were not painting what was in front of them. They were reaching for imitation, a scene of the picturesque, a view that married with the European tradition of epic, romantic, quaint, aspirational, meditative vista painting.

I've been doing the same. Grabbing for a vision devised elsewhere.

Pat and I agree to the live export contract. One hundred and ten of our heifers will be sent on a boat to China before the end of the year. Seventeen days at sea, and then a life elsewhere in a farm that resembles a feedlot.

The other option is to quit.

6.

Shortcuts

He used to stand at the sink in our one bedroom, 55-square-metre apartment, washing dishes. His broad shoulders slightly hunched with exhaustion from the strain of a shift working construction. Each gesture is careful, measured, and purposeful, so as to not break a glass or mug. Things purchased are precious. Now Pat's face, not this back, holds the exhaustion. Some mornings the brow will unfurl, and the day will be hopeful; it's rained, the stock agent has phoned the night before, and no animals have died.

'Photosynthesis is happening!' he'll shout jubilantly.

Other mornings will be darker. I can see it most clearly in his red-rimmed eyes that are either cloudy or glistening. It's said that women are invisible farmers, but it's less frequently acknowledged that farmers are invisible men. Stoicism is demanded so often, and if your heart is exposed and your mind rarely quiets, as is the case with my husband, then farming can be a torment.

He has been cutting gorse and broom in the steep back paddocks, ridding the farm of a pernicious weed that renders grazing land unviable if left too long. And it's been left too long. My dad, who farmed this lease block by himself before we arrived, did little maintenance. An accident decades ago damaged his back, reducing his capacity for hard physical labour. Fence wires curl loose around strainer posts, the paddocks choke with Scottish thistles, and our cattle stumble on discarded fence posts hidden deep within the grass.

My husband is building a tower of resentment towards the old guard as he labours, and as each hour passes, he sees his savings, invested in this farming business, go to waste.

Change requires a commitment to vulnerability. To not know what to do next. If you exist in a lineage of convention, a pattern of habits that year in year out stay the same, it's safe to keep going. The advice given enthusiastically by my dad and his ilk is staid and loud. I've stopped listening, but Pat hears it as a taunt given to belittle his work.

My ears are instead adjusting to the sounds of the land; the urban morass of auto noise, construction, and city living has departed and been replaced by the distinctive beat of tui wings as the white-necked bird soars, nectar-drunk, from one kowhai tree to the next. Walking beside a stream near the lake I hear the swish of a rainbow trout beating its way upstream long before I see its silver side glisten. I wish Pat could hear the tui wings beat or smell the faintest whiff of cocoa in the clod I pop up from the paddock after the last rain, but he doesn't stand still long enough.

Our daughter is cut in his image. As the quad bike rumbles distinctively down the hill track towards the house, her watchful gaze drifts to the glass sliding doors and she mouths 'dad'.

Some days I think he won't return.

The back paddocks are steep. The quad bike is notoriously unstable.

Some days he's not so careful.

For more than two years now, I've reached first for text to guide us away from our problems. I read excitedly about the vast, living networks within healthy soil where more species can be found than above ground in the entire Amazon rainforest.[1] I read about the wild grasses and tubers that once dominated the Australian landscape, and rewilding projects that transform an unproductive dairy farm into a low-lying forest in an attempt to diversify and bring back the birds, and encourage the natural system to flourish.

But I can't pull down the fences or turn this land into a rewilding oasis — not under the lease terms and not in our current financial predicament. I'd end up with land covered in second-growth bush, populated by turkeys, considering that New Zealand's endemic land species are predominantly birds. What I want most, though, is to move away from what rewilding advocate Isabella Tree describes as 'this conventional industrial chemical system of farming globally'.[2] But the answers to our problems aren't in books or on another's land — they're here. Where I write. They're in the distance between the pukeko's wetland territory and the parameter of the calf shed.

It's been more than two years since Pat and I arrived on this farm, and for the first time I am seeing its veins.

Pat comes home safely. We sit drinking coffee on the seat outside the sliding doors as our daughter crawls into his lap. 'Tell him not to come out anymore,' he says quietly. 'Tell him to stay at home. Just for a while.'

I need to stop moving; Pat needs room to move. Without judgement, criticism, or the weight of tradition. Behind the veil of COVID-19, we lock the farm gate and Pat starts anew. My dad, a symbol of a tradition that threatens Pat's identity and self, stays away. I didn't need to ask.

Small changes come first.

We agree never to rear calves again; instead, we'll buy them weaned from a cow. That decision alone lifts shadows from Pat. Next, he combines the cattle we have left on the farm into two mobs that balloon to 300 or more animals: a dancing river of cattle that rush through gates as they open into new pasture. The satisfied collective bellow drowns out the protest chorus of 'No. More. Cows.'

In the afternoons, he sets up temporary break fences in the large paddocks, sectioning the pasture into small blocks. The cattle move in, eat, and move out in quick succession, and leave behind them trampled plants and deposits of manure. As the weather grows cold again, and the grasses stop growing, giant rounds of silage and hay are unfurled and rolled down steep hills. The cattle crowd the flank of the feed and then use the remains as a warm bed. Then they move on to the next block. Over the months, the cattle round out and grow playful, and Pat's efforts are creating layer upon layer of organic matter on the land. Matter that will decompose and help create life just below the surface.

The tui have arrived back in the garden, but this time there are five of them. Small birds with colourful feathers appear, too, and stay. Some nesting in the paddocks, safe now from cropping machines and fertiliser spreaders. Some are common: fantail, finch, starling, sparrow, blackbirds, mynah, bellbirds, plovers, mallard and Paradise ducks. Others we've not seen before: a white-faced heron, a New Zealand falcon, goldfinch, greenfinch, and the rosellas — hordes of chattering Australian imports who move en masse, chasing and diving like kites in the wind.

One afternoon we move the cattle behind the house into the small tree stand that's home to the rosellas, and they turn and kick upon entrance, delirious with feed choice and obstacles to play around and scratch against. The paddocks pop with new growth much earlier than normal. Long-legged spiders wobble over the tall pasture; the apex predator in the insect world has arrived in the paddocks, which means there is life. Pat's new grazing system is working.

Graze a third, leave a third for next time, the other third to regenerate. It's the same principle for foraging, the same principle for kaitiakitanga, managing the growth cycle for the future. All those hours that he shut himself away in the bedroom, Pat was learning about a different way to farm: a style of farming that is gaining currency around the world. Elsewhere it's called regenerative agriculture; here it's Pat's Way. Smart farming, my uncle finally admits one day.

To follow a regenerative farming model is to admire the work of American organics pioneer Robert Rodale, and Allan Savory, the Anglo Zimbabwean ecologist who wrote the handbook for holistic management. Savory talks of a transformation from near desertification to lush habitat through the interconnectedness of grazing animals and plant diversity. But he also talks about a transformation in life: of talking to each other more, deciding on what values we cherish most, and working to insert that value into each action and plan.

Initially, I was sceptical. His model of land and animal management grew out of an environment where large ruminant animals have long existed: the plains and savannas of North America and Africa. I was uncertain his methods would work, or were needed, on land where herds have only ever been introduced, like in Australia and New Zealand. I now understand, though, that a farmer's job is to have the humility to respect that complex adaptive systems can never be fully understood. All we can do is work within parameters to empower them.

Yet where we see potential, others see threat. Regenerative agriculture's fashion has inflamed those wedded to conventional farming systems. 'The result is a decrease in pasture quality, a decrease in efficiency of production, a decrease in organic matter,' cries a former chief scientist of so-called regenerative practices. Another blasts that 'Soils need fertilisers to be productive.' Yet another describes regenerative practices as scientifically untenable, bordering on mythology.

But the critical voices didn't see the worry in my husband's eyes, the change in his posture as the weight of financial failure hammered against his shoulders. They can criticise, but it doesn't mean they're right or that we'll change back. Regenerative farming is not prescriptive. There are grounding principles — like keep the soil covered, minimise disturbance to plants' roots, increase diversity, and manage livestock to promote healthy soil — but the most important point is to *know your context*. It's the advice I've been looking for most. It's so simple. Know your context, and act. What objectors see is a threat to the portion of the hourglass system that has a vice-like grip on the global food and farming industry. This is farming that threatens input companies — those that make, sell, and guide farmers towards fertilisers, pesticides, genetically modified seeds, herbicides, heavy machinery, winter cropping, ryegrass replanting, ploughing, and tilling. And we are phasing it all out.

What is gained is time, birdsong, worms, and fungi. And a life.

Farmers are grumpy by lore, but statistically that picture is bleaker: farmers are killing themselves. A 2020 report by a cohort of medical and public health researchers found that Australia's farmers are an at-risk group for heightened rates of suicide. And the reasons are myriad: poor access to support services, unsustainable work ethic, uncertainty and lack of control in farming, social disconnection, poor business profitability, acclimatisation to risk taking.[3]

One young farmer I meet has turned away from conventional practices in part because he watched those using them sag under the weight of mental ill health. He told me the attraction to regenerative farming is that 'it's got a totally different spin on it'. It's positive where for so long the tone has been overwhelmingly negative. And angry.

The optimism that creeps back into our home is also about money, as it always is.

Tim, a high-country farmer who helped us early on to understand what regenerative farming means in technical terms, told me his tipping point was the realisation that 'We'll never get ahead when we're at the mercy of the market.'

His family farm 5,500 hectares in one of the driest parts of New Zealand. On a good year, they might get 450 millimetres of rain. It's a moonscape in summer, dry high hills potted with rocks spat long ago from a volcanic eruption. Split with a vein of brittle azure water that is the Clutha River. All Tim can do in the ever hotter, drier summer is to focus on carbon. The storing of carbon in the soil. Carbon can hold its weight in water by more than ten times, he tells me as we sit at the kitchen table in his refurbished workers cottage with the old native-timber floorboards. 'Every percentage of carbon I can get in the soil,' he says, 'can hold 30 tonnes more water.' That's how we'll get through.

In the small gestures of unconventional farming, we untether from the system that's driven by the old guard and the industrial farming industry, which prioritises and celebrates high yield, fast growth, and massive inputs. Our cattle and a bit of seaweed tonic, not the spreading truck with its tonnes of synthetic fertiliser, are now helping to bring life back into the soil — and that realisation runs counter to the headlines that keep telling me cattle farming, in its current form, globally, is an unsustainable polluter.

We pulled the cattle deep into our value system — environment first. They, too, must work for the land, water, and air; that is the job, the labour, the life.

Now, in the detail of the climate reports, in the lines that harried journalists and op-ed writers didn't read, I see the proposed framework for the great food transformation as it was intended: an acknowledgement that local adaptation is needed to shift to the desired sustainable food systems. It's not a call to create division — it's a call to know your context.

In smaller numbers, managed a certain way, our cattle, our methane-emitting environmental degraders are useful, here, in this place. But not everywhere. Know your context, progressive farmers tell us. And … be brave enough to change.

Emeritus Professor Frank Griffin, the director of Otago University's agriculture research unit, Ag at Otago, assured me once that: 'In the background, farmers are the most incredibly creative and sensitive bunch of people.' I see this in Tim's gestures as he talks about soil that smells like chocolate cake and of learning to read the earth like a braille text. Weathered, in his forties, with cracked hands and eyes squinting against the sun, he sees a need for more organic matter, and reaches for a colourful cocktail of legume seeds, sunflowers, clovers, and peas to regenerate his land after decades of heavy inputs. But not all share his vision; others need a different incentive to chase carbon.

Money. Yes, that again.

In 2012, Wilmot Cattle Company started measuring soil carbon and soon afterwards implemented a fundamental shift towards intensive rotational grazing — a key part of regenerative farming. The result, they found, was a soil carbon increase from 2.5 per cent to 5 per cent. With backing from the farm's investors, Macdoch Group, they used their detailed multi-year data trove to secure a private market contract via the US-based carbon trading company Regen Network. 'We don't consider ourselves soil farmers,' says the farm manager Stuart Austin.[4] The goal is to keep producing quality beef off land that's thriving despite the hotter, drier world. They're farmers who've faced up to climate change and received a bonus because they've monetised the carbon and on-sold 43,000 soil carbon credits to Microsoft under a 25-year deal worth A\$500,000. Across 2,700 hectares, 43,000 tonnes of CO_2 are sequestered. And at approximately A\$11.60 per credit, soil farming is a bonus worth pursuing.

Included in the Australian federal government's National Soil Strategy is a soil market focused on 25- and 100-year soil carbon projects, similar to what Stuart has signed up to with Regen Network, but with a measurement system viewed as the gold standard. The goal is to encourage farmers to value soil, the lifeblood of the country, and to prioritise it over and above all else.

Some in the scientific community remain cautious, unconvinced that soil carbon sinks can truly help the world hit its Paris Agreement targets. The calculations for measuring a baseline in soil organic carbon (SOC) and then improvements are wavy, and there is little global consensus over the quality, quantity, and duration of SOC data capture needed before credits are allocated by verification bodies and sold on the market. Paige Stanley, an American soil researcher focused on regenerative grazing practices, fears verification bodies have put the market before the science, selling shoddy carbon credits based on little testing.[5]

Some also fear that if trading relies on increasing carbon, there exists a potential financial risk for farmers who lose soil carbon through increasing flood and drought — in other words, through the consequences of climate change.

Stuart tells me Wilmot has planned for this and reserves 25 per cent of earnings in case carbon losses require a reimbursement. It's good accounting. But he's confident it won't be needed. The farm went through the recent dry period without losing a percentage of carbon. That's the point: increasing carbon means the entire farm system improves.

In the US, the California Department of Food and Agriculture is issuing payments to farmers for their carbon stores if they opt for cover cropping and no-till or reduced-till practices — techniques for planting crops embraced by regenerative advocates.[6] The Biden Administration is reportedly aiming to replicate the scheme in order to create a carbon sink across the country's vast farmland, part of the effort to cut the country's carbon emissions by 50 to 52 per cent below 2005 levels by the year 2030.[7]

Companies like the US-based start-up Nori are offering individuals the chance to further scrub clean their carbon footprint by buying credits sourced from registered farmers. It's a company Edward Towers, a progressive young dairy farmer from the Lune Valley in Lancaster, UK, has his eye on. Edward is studying the potential of creating carbon-linked cryptocurrencies with the aim of accelerating soil improvements.

He tells me the appeal of cryptocurrency is due to the use of blockchain, which can provide useful management tools for a carbon sequestration supply chain. He points out that anyone with access to the internet can trade and circumvent government control, which potentially makes it a useful tool to take 'global and tangible steps toward rapid climate action'.

Edward is a fourteenth-generation Towers farmer. He's not a cryptocurrency expert, but he's committed to pursuing this research out of 'regret minimisation'; he doesn't want to be a grumpy old farmer who looks back and thinks 'I should have done more'. His brother, Joe, farms alongside Ed at their family property in the Lune Valley. They sell milk, some commercially, some under their own brand, Brades Farm, direct to select cafes as barista milk. They've worked hard to build a connection with their consumers to repair a broken relationship between farmer and community. But Joe is concerned that too few are willing to follow suit, opting instead to 'stick to their own'. Farmers, he says, are 'a little bit scared to mingle with city folk because of the way that we feel'. They feel like they're misunderstood. They feel like their identity as farmers and food producers is near worthless because the UK has backed cheap food from elsewhere.

'I feel like the enemy is people who don't care where their food comes from,' says Joe, before clarifying that in this opinion he's speaking from the heart, not as a spokesperson for the family company, as he is charged to do most days. This slippage reveals a deep sadness that the system isn't working for the farmers or the consumers. That the divide has become war-like.[8]

Here, where I am, it feels like the split is rupturing the farming community as well. Those who are experimenting with regenerative practices often do it in the back paddock, away from the glare of neighbour judgement. One brother doesn't tell the other brother who farms next door that he's developed a particular interest in soil. Regen brother wants to get through another long, hot, dry summer without losing pasture and stock. But he stays silent about his experiments with diverse cover crops and seaweed fertiliser for fear of failure, blowback, and ridicule from the brother who subscribes to the view that New Zealand's farming sector, with its grass-focused, 365-day outdoor system, is the most efficient and environmentally sound production sector in the world. *We don't need to do more*, brother says. *I'm not degrading the soil, so why do I need to regenerate it?*

New Zealand soil holds on average 90 tonnes of carbon per hectare in the top 30 cm of soils.[9] Researchers estimate that if the top 0.3–0.4 m of agricultural land could sequester 2–3 gigatonne (Gt) of carbon per year as a result of soils remaining untilled with plants in place, globally this would effectively offset 20–35 per cent of all human-made greenhouse emissions. That's an enormous amount of carbon mass. Consider this equivalent — just one gigatonne is a mass roughly twice the size of all humans on earth combined. Surely that's a good reason to get excited about soil? After all, the goal is a cooler, livable planet.

In the early morning Pat stands at the door, looking out over the house paddock. His face becomes a palette of wonder. 'Nature actually works like the books says,' he says. If you stop trying to interfere.

Work with nature, not against it, is the fundamental ethos of regenerative farming. And it works for us, and yet it's not the norm. Why? Because technology has been promised: agricultural technology that will drop a farm's methane emissions and allow them to keep operating as they have been — high yield, high production, high inputs.

For some years, each May as the southern winds started to chill the coastline, my mum would take my sisters and me to a beach village called Waimarama on the east coast of New Zealand. Few holidayed at the beach in winter, but we did.

My mum would set off on long lonely walks down the wide stretch of hard wet beach, bringing back ocean trinkets to stack on the table or arrange in a vase: paua (abalone) shells rubbed luminous and blue on the inside by the cleansing ocean, petrified wood brittle and abstract, pipi and cockle shells, and oddly coloured rocks. We'd rush into the ocean up to our knees, squealing and then retreating. Few swim in May.

On an overcast day we'd hunt for seaweed, walking kilometres down the beach and over the rocks at the southern end as Te Motu-o-Kura Bare Island loomed like a grey wave above us. I would pop juicy pods of Neptune's necklace between my fingers and drag long trains of blackened weed, leaving track marks of a strange creature on the beach behind.

Now, that beach holiday plaything preoccupies me. Seaweed, I'm told, is the golden ribbon of the ocean, designed to heal our collective wounds. It's a food source, a methane suppressor, a natural fertiliser, a vital tap of nutrients, an ocean cleanser. Something that will solve a big part of the emissions puzzle for food producers. Much more than a toy to be dragged along an empty beach in the depths of a New Zealand winter.

The dairy company Fonterra has announced it is partnering with universities, the Australian government's CSIRO, and the Australian company Sea Forest, a start-up with ambitions to cultivate a 'natural solution to climate change'. The intention is to grow a variety of seaweed called asparagopsis, endemic to the cold waters of south Australia and New Zealand, at a scale that will allow for regular harvest and processing. This would turn it into a supplement that early studies from CSIRO suggest can reduce methane emissions in dairy and feedlot cattle by more than 80 per cent.[10]

The dairy industry holds a peculiar place in New Zealand's farming lore and economy. Cows were brought out by British settlers, and the first export of dairy was sent from the South Island's Banks Peninsula in 1846. Since then, the industry has been tethered to the government, no matter what side of politics is governing. The Dairy Board's status until the mid-1980s as a tax-exempt, single-selling monopoly cemented its power.[11]

When I was a kid, a dairy farm would probably be around 70 hectares, owned by a family who would milk 140 cows, or thereabouts.[12] Now, just one dairy farming conglomerate, buttressed by investments from the likes of Canadian pension fund Sooke Investments, owns and milks more than 50,000 cows. The national herd numbers just under 5 million cows. Dairy is too big to fail in New Zealand, and that often guards it from the glare of the moral spotlight. Environmental and animal welfare concerns are raised, but no one is asking if we *should* be farming this way, because the simple answer is that we *can* — so we do.

In New Zealand, the specific target is to cut biogenic methane emissions (from ruminant animals) by 10 per cent of 2017 levels by 2030. Ask industry leaders how this will be achieved and a number, like Federated Farmers president Andrew Hoggard, start talking rhapsodically about methane-suppressing cattle-feed technology. Let's not fixate on dropping herd numbers, is the subtext here. Let's not fixate on full system change. Hoggard reasons that if New

Zealand drops the number of cattle dramatically, a less-efficient higher-polluting country will pick up the slack, and that's not solving the global issue. 'There's good options out there, like a feed inhibitor option being trialled right now,' he told me to further his point about agricultural technology being the answer.

This vision, that technology will lead farming through climate change, is concerning to some who believe that the relentless pursuit of improved production will shoot farming straight past the land's natural environmental limits.[13] Back in 2004, Dr Morgan Williams, New Zealand's Parliamentary Commissioner for the Environment, urged the agriculture sector to pull back from intensification in favour of more sustainable systems. He saw the trend as a real threat to the natural capital. But the pace of change was glacial.

To understand why, Dr Williams suggested I look at the entire food production system to gain insight into the risks that farmers shoulder. Some risk is manageable, but, increasingly, the collective pressures of a changing climate and market demand farmers completely change how they farm. It's a leap many aren't willing to take, especially on top of all the challenges they're already facing. In Dr Williams' experience, a farm is much more resilient — both ecologically and economically — when farmers put aside the relentless pursuit of productivity and appreciate the value of 'enoughness'.

It's a theory that rings true beyond farming. Is one overseas holiday a year enough? One summer dress? One birthday present for your two-year-old? To define what is enough is a protest spear thrown at the heart of the indulgent consumerism that keeps global capitalism charging upwards in relentless unlimited growth.

When is it enough? What is enough? Others aren't even looking at this equation, they're just waiting on methane supplements so they can capitalise on high returns for milk powder. Up production, it's never quite enough.

We have far fewer cattle on the farm this year than in previous years when my dad would take on a mob of young dairy heifers on behalf of dairy farmers who didn't want to manage their own young stock. Like all things in the food system, we're in a trade-off. We've downsized on the farm to better manage the land, to improve the soil, and to bring life back into the waterways, but the consequence is participating in an industry many protest, because we still need profit. The heifers we've committed to the live export trade are still on the farm. A bull has been sent by the exporters; the heifers are expected to be in calf when they arrive in China. It's part of the deal.

We have a long way to go.

Sea Forest's chief science advisor, Dr Jeffrey Wright, tells me the methane-suppressing seaweed science is real and the results are there; they just need to keep going. They're doing careful experiments in labs to learn more about the seaweed's growth cycle and peculiarities, and I am in awe of his team's patience, given the urgency.

The company was founded by Sam Elsom, an avid surfer and former fashion industry creative turned climate campaigner, and early investors include pro-surfer Mick Fanning. Sea Forest, along with a small number of competitors, is pushing to commercialise land-based and deep ocean seaweed production based on the intel developed by CSIRO's FutureFeed project.[14]

In September 2020, CH4, another seaweed company, announced it was expanding its ocean-based trial off the brittle coast of Stewart Island with the aid of the NZ$4.45 million raised through a seed-funding round. The company's co-founder, Steve Meller, told me the company has clients lined up in Australia, New Zealand, and the US.

The deep-ocean growing system is purposeful. CH4 are in talks with fisheries companies to grow the seaweed on consignment around salmon farms. Nick Gerritsen, Meller's co-founder, is pitching that the seaweed plots will absorb excess nitrogen waste, a by-product of fin-fish farms, allowing aquaculture companies to scale up sustainably.[15]

But should we be content with sustainable or strive for something more? Fisher Claire Edwards told me the philosophy she subscribes to is to 'leave things better off than we found it'. She and her partner, Troy Bramley, run Tora Collective, a family-run fishery that holds a rare permit to harvest bull kelp from the east coast of New Zealand. They also hold licenses to fish off the coast for paua, kina, and crayfish. To sustain is to manage the status quo, she says. To regenerate is to future-proof for the generations yet to come. They want to improve the fishing industry, 'otherwise it just feels like we're capitalising on something. You've got to give back'.

Troy's parents made a life for their family on a rugged coastline in a small cabin without luxury; their playground was the land around them, their supermarket the ocean and beach before them. Claire and Troy are returning to this in spirit; they want to provide food that is tied to a specific place and season. But to do so is a small act of revolution: crayfish, like beef and dairy, are typically exported, tied up in lucrative licenses, and considered a luxury item of any table. Yet they need not be. At the heart of Tora Collective is a push for food for locals first. Claire wants to keep the kiamoana (seafood) here, and sell them to New Zealanders who can celebrate food and season with a fresh catch.

I was hoping to buy kelp noodles from Claire, but they're not selling seaweed, yet. One day she hopes to harvest beach-cast seaweed for others to enjoy, but for now Claire continues to learn from Troy's mum, the keeper of the harvesting knowledge. Instead, I buy a sack of southern clams from the supermarket. At $5 a kilogram they are a cheap protein, reflective of this place — an island nation in the middle of the South Pacific.

I sauté diced onion and garlic, add the cleaned clams in their creamy shells, flecked with green and pink, and a dash of cooking water from the spaghetti bubbling next to the pan. And cover. Five minutes later, the shellfish have split open in a cheerful welcome. In goes a knob of butter, chilli flakes, and fresh chopping parsley. It's served with spaghetti. Simple. Sweet. Nutritious. Affordable and of this place.

The process of seeking permission and waiting for the knowledge to germinate, as Claire is doing with the kelp, is antithetical to the global push for more fast, convenient food, more protein, more innovation to solve global warming. CH4 founder Steve Meller and Claire Edwards are at opposite poles but pushing towards a common middle ground.

When I speak with Meller about his seaweed venture, he is in Silicon Valley, California. Meller amassed a swag of degrees (B.Sc., B.Sc. Hons, Ph.D. in Physiology specialising in Neuroanatomy/ Neuroscience) at the University of South Australia and then made his home in the United States, working first for universities and then for consumer manufacturing giant Procter & Gamble, where he held senior executive positions in research and development. The pitch is that if approximately 100 grams of seaweed supplement, containing no less than six milligrams of a compound called bromoform, are fed to cattle per day, the result is a drop of 80–95 per cent in methane emission via burps and farts. Sounds simple, but there's a glitch.

Asparagopsis is a red seaweed species that contains a carbon compound known as bromoform in cellular glands. The trick is for the bromoform to stay contained in the gland during processing and then released in the cattle beast's rumen system — the stomachs — while the fermentation of feed occurs. It doesn't prevent the fermentation process but rather suppresses the gases released as the bacteria goes to work on hard-to-digest grasses and grains.

Objectors worry that if bromoform is released into the atmosphere en masse it will have a degrading effect on the ozone layer. If the technology isn't right during processing, this simple solution to warming could do more harm than good. But if it is successful, then 5,000 hectares off the coast of South Australia can produce enough asparagopsis to feed the 2.5 million dairy and feedlot cattle in Australia. And at scale, Meller claims the asparagopsis product can fill the shortfall between what governments committed to under the Paris Agreement and what's required to keep warming below two degrees in the next decade. This is a trade-off — focusing just on methane in the next decade can potentially lead to cooling but it's a short-term fix. Projections suggest cattle will require 100 g a day, ongoing. It's not a one-off — it's a daily feed for dairy cows and a daily supplement for feedlot cattle.

What about all the free-range cattle? 'Not my concern,' says Meller.

His is a methane solution for intensified cattle farming the world over; the type of farming that people hate to see but which is on the rise because, as far as the numbers go, it's an efficient and low-impact way to produce beef. Meller told me the rush of interest from meat processors and retailers in the seaweed supplement is because they're pushing for 'low carbon' or 'zero carbon' labelling. They want to present a feel-good purchase option for conscious consumers who prefer beef steak to lab-grown or pea protein, but that beef, or milk, if it's claiming carbon neutrality based on the cattle consuming methane suppressants, is most likely coming from an intensive dairy or feedlot system where the feed can be monitored. So, the question remains: if we're to continue to eat beef and dairy and strive for a zero emissions industry, are cattle destined for more intensive farming systems to get their daily fix of seaweed?

Is that the vision conjured when you pick up the pack of 'zero-carbon' beef?

What is clear is that those who graze cattle on pasture, like we do, need to find another way. Do we go backwards towards a style of farming and eating that existed before the wave of technology allowed us to do so much more?

I've heard it said by critics many times that regenerative agriculture is 'just like my grandfather did it': farming without the technology that science provides. And yet these same commentators baulk at meat grown in labs or dairy cultivated in Petri dishes. When is technology good, and when should it be avoided? Is tech phobia just another bias? Because, really, what's so wrong with mince that doesn't come from an animal if it's affordable to those who can't afford the luxury of shopping for carbon-zero beef and dairy, and it's proven to have a lower footprint?

As Michael Carolan told me, there's no such thing as a food without impact — we just need to choose what we value most. There is a group of like-minded farmers and food entrepreneurs who value the same thing as us above all else: the environment. But they remain poles apart, because when you're trying to sell a new product it's useful to have an enemy, something that's definitely worse than what you're selling. A target to shoot one's marketing campaign at or past.

In 2019, Impossible Foods, the San Francisco food manufacturer that makes plant-based proteins like a bleeding veggie burger and has an ambition to eliminate the need for animals in the global food system by 2035, issued a rebuttal to Allan Savoy's public statements on the potential of regenerative agriculture.

The Impossible Foods '2019 Impact Report' makes several valid points: 'Iterating on animal farming will not stop the climate crisis'. True. 'No rancher or farmer wants to damage the land they manage'. True again. And it goes on: 'Sure, regenerative grazing may beat industrial bovine strip mining [feedlots] on a couple of fronts — but it's all still rooted in the same inefficiency of animal metabolism.' The solution: plant-based technologies, a food-making process that provides the means 'to reverse the whole unsustainable system'.[16] Despite our efforts to farm this land well, we are still someone's enemy in the fight for the planet (and profit).

The Good Food Institute in San Francisco is in the business of promoting and advocating for plant-based food products and start-ups like Impossible Foods. Their communications manager told me the group is actively engaging with ranchers across the United States to help the willing manage the transition from ranching to growing because plant-based burgers require vast quantities of isolated ingredients from crops including corn, barley, wheat, soy, and pea.

But ask a rancher, like Nicolette Hahn Niman — a vegetarian lawyer turned author and meat advocate who farms in Northern California — if this move away from cattle grazing is good for the land, and you'll get a counter approximating this: 'Alternative meats are based on the monoculture, chemical-based, fossil fuel–based, ecologically devastating agriculture system that is in place right now, which desperately needs to be undone.'[17]

Hahn Niman recalls driving through the state of Idaho, seeing acre after acre of potato fields, and then through Iowa, where the view is the same, but the crops are soy or corn. 'It literally looks like the moon,' she says.[18] It's a barren landscape where monocrops are rotated and fertiliser is dropped season after season. The greatest irony, Hahn Niman says, is that this is the solution being proposed to combat climate change and land degradation.

Some are in the business of marketing; others will farm to produce food. Impossible Foods is exemplary at the former. The company has publicly stated that its consumer growth strategy is based on a campaign to brand animal agriculture as a huge contributor to climate change. It's not enough to simply market the product — they need a target to hit.[19]

Josh Tetrick, the CEO of another plant-based start-up, Just Inc, says plant-based technology exists because people are imperfect. They may have a desire to do good, but how to do good is too complicated. So Just Inc provides an easy solution: buy mung bean–based 'eggs' or the company's lab-grown clean meat products, once it's on the market in countries other than Singapore. Then you can be assured that the footprint of your food is delicate, according to the company literature. Tetrick projects an image of the socially conscious food entrepreneur; he worries about the health problems caused by processed foods, and the fact that 'about a billion people are going to bed hungry'.[20] He considers the animal agriculture sector to be a massive driver of climate change because, globally, cattle ranching and farming is a driver for forest clearances and burning, and it encourages farmers to grow crops for feed rather than food. 'We have a food system that is not rational,' says Tetrick in his Alabamian drawl. 'It's not fair. It's not equitable. I don't think it tastes good enough. And we need to fix it.'

Agreed. We need to fix it. But the how is where we come undone.

If the conversation remains stuck on cattle burps, Meatless Mondays, and emission rates, we face inertia and failure. That should frighten everyone.

'Food tech shouldn't be necessary,' says Tetrick. He would prefer that people ate whole foods like apples, kale, spinach, collard greens, grapes, cauliflower, but for the most part they don't. So, he wants to make it a little bit easier for people to eat well and do their bit for the planet without needing to grapple with the nuances embedded within the science of global warming.

The goal at Just Inc is to get the price of the products below that of their animal counterparts, to remove the main barrier to change — which is cost. If they fail, 'Well, you're not gonna have the kind of transformative change that you want,' says Tetrick.[21]

Unlike some of his colleagues in the plant-based production sector, Tetrick is not lobbying for the destruction of animal agriculture: he prefers to partner with Big Ag. To distribute his mung bean eggs efficiently, he works with egg distributors like Italian producer and distributor Eurovo Group and Post Holding's Michael Foods, the largest distributor of eggs to restaurants in the US. And in his efforts to make high-end lab-grown beef he's partnered with Japanese Wagyu producer Toriyama to source cells from which to grow his clean meat.

Tetrick is working to shift the foodscape from inside the juggernaut, effectively cannibalising his host. After all, to follow a similar marketing strategy to Impossible Burger et al. would be disingenuous, because despite efforts to relabel lab-grown meat 'Clean Meat', denoting a slaughter-free process, the product still relies on an animal product, namely Fetal Bovine Serum (FBS) — a nutrient-rich cocktail sourced from a bovine fetus. It can be harvested without slaughter, but it's typically sourced as part of the abattoir process.

The fledgling clean-meat sector is working to engineer a synthetic equivalent to FBS, but for now it's not completely clean. Nor is it less environmentally damaging than conventional livestock farming if the manufacturing facilities continue to rely on fossil fuels to power production and distribution.[22] But there is a glint of hope in the endeavour: if renewable energy replaces fossil fuels, the footprint is projected to be lower. If the cost can compete with cheap mincemeat, the potential to reduce the global herd, and therefore methane emissions, is real.

The same potential exists in labs for dairy products.

Perfect Day, an American company that produces fermented animal-free whey protein for use in manufacturing, markets itself as 'Kinder, Greener Dairy'.[23] In the company's first independently reviewed Life Cycle Assessment, which aimed to measure its product's environmental impact from cradle to factory gate, it concluded that Perfect Day whey protein is up to 96.6 per cent lower in greenhouse gas emissions compared to the protein derived from cows.[24] A comparison made after reviewing five meta-analysis studies surveying LCAs from dairy farms around the world. Ask a New Zealand dairy expert how they fare, and you'll receive the response that New Zealand is the most efficient producer of fat- and protein-corrected milk (FPCM) — an equation the International Dairy Federation uses as a global standard to measure the carbon footprints against the nutritional value of milk product. In New Zealand that figure is 0.74 kilogram CO2e per kilogram FPCM, which is, according to dairy executives, 46 per cent less than the average of the countries studied.[25]

We can be dazed by these numbers, relying on interpreters to guide us through, but the salient point is that technology proven to help drop emission rates should not be feared, because every part of the food production chain needs to shift. The status quo got us to 1.1 degree warming; continuing as is won't keep us below 1.5 degrees. If the agriculture and food sectors achieve a rapid reduction in emissions, the effect is cooling:[26] an actual drop in temperature, which presents an opportunity for agriculture to compensate for delays in reducing carbon dioxide emissions.[27] And make no mistake, the fossil fuel industry is trying, has been trying, to sow doubt over the need to change for decades.[28] It's doing all it can to retain the status quo. Farming has the chance to be the short-term saviour, rather than the culprit, but only if we stop throwing mud at each other.

I stand in the paddock in front of the farmhouse, facing north against the wind that rushes in over the small lake, ruffling the downy grey feathers of the cygnets as it passes, and I scream: 'We need to get to zero!'

Farm

Everyone needs to act!
Now.
But I'm screaming at swans; no one else seems to be listening.

7.

Change maker

Working with cattle is a peculiar pleasure. They're curious and playful animals that easily form a type of domesticated bond with humans who handle them often. Some stockpeople choose to assert their dominance over animals, to manage them in a fury of bikes, horses, dogs, and ropes. I work with my young cattle on foot, with a dog named Red who zigzags behind the herd, nudging them along. It's slow and mostly calm. And it's pleasurable.

The hedonist in me wants to walk slowly behind my 300 cattle, moving them from grass-covered field to the next, then return home to eat a meal of slow-cooked beef stewed in a heady cocktail of stock, tomatoes, and spices including paprika, cardamon, cloves, coriander, fenugreek, black pepper, allspice, nutmeg, and chilli. What pleasure.

The environmentalist in me awakes that same night with indigestion and night sweats from fear that I'm destroying my part of the planet faster than I can bank the meagre profit we make selling 100-kilogram bull calves.

Are we right to use cattle this way? Purely for our own means?

—

At the height of summer, the grasses are knee-high and scratchy against the bare legs of a six-year-old. When I was little, my sisters and I would tear through the blades with a piece of woven bale, flagging ewes and lambs into the open mouth of a temporary yard gate. Some of my happiest childhood memories are of the days we spent as a family in hot paddocks docking lambs.

I spent the first part of my childhood on a sheep and beef farm in the middle of the North Island of New Zealand. And we all worked at farming in some small way. This was when my parents were still married. My mum would pack a picnic lunch and a thermos of black tea with milk and a little sugar. It was all delicious in my mind. The soft warmth of the midday sun, the bright light of summer, the smell of dry grass, the sound of lambs bleating, a mouthful of sweet tea, and a crumbly scone with creamy butter. The violence of what we were doing out there was well hidden under the veil of normal farm life.

Docking is the process of affixing a small rubber band around the tails of lambs while they're still small enough to hoist into a cradle and lie supine. Over the following weeks the blood will stop flowing to the tail and it will fall off. The removal of the tail makes it easier to prevent fatal fly strike — a clean term that describes flies laying eggs that hatch in faeces-covered tails, resulting in maggots that slowly eat into the flesh of the animal. So, farmers in this part of the world dock; it's done in the aid of animal welfare, but also, and more importantly, the production of quality meat.

These things, acts that are at the core of the business of Southern Hemisphere farming, were normalised for me when I was a child. I always wanted to sit next to my mum and the picnic hamper as she drove the potholed dirt tracks heading towards the back of our farm and the temporary yard full of sheep. At night, I would hungrily pull succulent lamb meat from the bone, licking the fat from my fingers afterward. But a lot of what was done then is now banned or regulated under new laws. Social mores, farming norms change. That's how it should be, and the glow of childhood memories has long since faded.

The live export heifers were scheduled to leave the farm months ago, before the end of the year. But they remain. First, we were told it was because the ship was delayed arriving. COVID-19 had clogged the global shipping network, and that included live export boats. Then we were told nothing. 'Soon,' they said. 'Soon.' Between the first and second 'soon', the heifers were tested for disease and their frames inspected for any 'defects' in breeding. All passed. Then we were told half of the mob were too heavy. We'd been contracted to deliver them at a certain weight, and they'd gained much more than that. A point of pride turned into a contract negotiation. What were we expected to do? Ration their food so that they stopped growing? Yes, was the answer.

For months this went on, and we started to draft the heavier animals out of the mob. We'll keep them, I said. They'll calf here. It will be the start of something good out of a terrible decision.

Finally, four months after the contract deadline, a truck arrives to collect the animals. They will leave for China after a period in a quarantine farm, adjusting to the feed they'll consume on the ship. Just like at Tess Herbert's feedlot, they'll be given high-energy grains. We are sending them to a feedlot. No matter the money, now far less with delays — the bridge between us continuing to farm cattle and the food we hoped to produce is a level of suffering and death for the animals.

That's farming, some say, flippantly. Others, like the New Zealand Labour government say *no, we can be better than that*. Mere weeks after the heifers leave the farm, the government announces a ban on live cattle export by sea, effective 2023. Continuing the trade, according to the Agriculture Minister Damian O'Conner, presents an 'unacceptable risk to New Zealand's reputation'.

We did what we did for the money, and who is to say what the damage is to our reputation? We await judgement. From people we know — our city friends and other farmers who deplore the trade — and those we don't, those who are looking for yet more reasons to rail against cattle farmers. I know that to ignore an animal's looming death and sanction the suffering is to reduce the animal to an object. And an animal is not that. It is far from that.

As winter dragged, the pregnant heifers started rounding in the belly and grew full in the udder. In late October, a tiny black calf appeared one morning, born without trouble the night before. The group headbutted and nudged the tiny specimen around in circles as the heifer tried in vain to shield it from one bully in particular — a head-high animal with a white face, black body, and black eye patches; she was intent on causing the others harm in her enlarged state. Another chose to run defence for the first born, and she trailed the new mum and calf around the paddock, letting them feed in peace. She was a shadow, and three weeks later she gave birth to an enormous bull and became the matriarch of the mob.

The bully produced a stillborn calf and stood bellowing over the spot for hours: a cry familiar to any animal. Then mastitis developed. Our vet arrived to administer treatment for the infected teat and the headstrong heifer became meek, standing quietly as the vet worked to ease her pain.

We don't purposefully humanise our animals. I watch with confusion when dairy cows are named, cuddled, ridden, coddled, and taught to urinate on a specific spot.[1] To ascribe layers of human attributes to cattle is to squash their natural tendencies; it's just another type of dominance.

They are not human, but nor are they objects.

This became so evident watching the heifers form maternal bonds with their offspring and each other. As the remainder of the mob calved easily, a routine was established. Each day one watched over the calves as they slept, curling against the hard spring wind, while the others grazed. A creche, of sorts, was formed. At each subsequent birth, the matriarch would be first to greet the calf, nudging the others out of the way for a moment.

The reading of these gestures is fodder for researchers interested in animal welfare. It can also be ammunition for those who believe these animals should never be killed. There is evidence of social and physical cognition; any good farmer will tell you cattle form grazing habits based on memory. They know our farm as intimately as we do, gravitating to the best feed spots as soon as they're in a paddock, all the while looking for the plants that will give them the tonic they crave. And while researchers note a lack of evidence that cattle can manipulate objects for an end goal, I can assure you that a cow can open a gate with her tongue to reach better feed on the other side.[2] But there is also something deep inside that remains wild. And that is to be respected, always.

We learned this painfully, when we drafted off a group of white Charolais heifers for sale and drove them into the yards to weigh. They were never quiet, even as calves, but they weren't dangerous. Yet on this hot afternoon, they broke fences by kicking back with their hind legs, jumped gates, ripping out hinges, and launched themselves out of the weighing chute.

Never turn your back. Never underestimate an animal. On calm peaceful days, I forget that for hundreds of years humans have worked to domesticate animals for their convenience, some more successfully than others. But deep within remains a fight or flight response, which doesn't diminish in us or in them.

The fight I choose is the same as that of the cattle breaking wooden boards with their hooves; it's survival. But mine is a longer campaign to see my daughter, and perhaps her daughter, live without the fear of raging and recurrent storms, floods, heatwaves, fires, and droughts. The campaign to cool the planet requires a sacrifice, from them, me, and all of us.

When I started digging for worms on our farm, I popped clod after clod of tight, knitted grasses covering little more than dust. Now, because we've changed the way we farm here, the way we manage the animals and how they graze, the clods look different. Each spade contains lumpy soil bound together by nodules and deep roots, and worm tracks. Worms. So many of them. While others are chasing profit, I'm chasing life, and so much of it is barely visible to the naked eye. It's all life: us, them, the community we sell food to, everything that's thriving in the little ecosystem we've created. But it's a fine balance. If we push the land too hard, cracks will form again. And so, sacrifice is demanded —the cattle must be gone from the land before they grow to a weight that crushes these pumice soils, sending sediment and slips into the stream below. Gone before the animal's appetite outweighs what we can grow naturally. Gone so that nutritious food, high in protein, iron, zinc, and vitamin B12 is circulated into our community.

Does that justify my work here, the killing of animals for food?

There is a philosopher named Jeff McMahan who holds the position of White's Professor of Moral Philosophy at England's University of Oxford. And, surprisingly, we have a bit in common. As a child growing up in the American South, he spent time shooting ducks and 'birds of that sort'. He was given a shotgun at the age of 12 and taught how to hunt. But at the age of 16 he came to believe that what he was doing was wrong. At 17 he became a vegetarian.

I spoke to Professor McMahan over Skype one evening. It was morning for him, and he was busy dealing with house repairs. Thinking back to his childhood Professor McMahan says: 'I remember seeing photographs taken by friends of mine in which they and their fathers were posed in front of perhaps 50 or 60 ducks that they had shot one morning.' They couldn't eat that number of ducks. 'They would shoot them and then just leave them to rot.'

There is a wood-framed black-and-white photo on the wall of my father's house, and in the foreground are rows of dead ducks arranged neatly on the grass. My father is a boy in the photo, around 11 years old, posing with his father. He has a lopsided grin on his face and one small hand grasped around the barrel of a shotgun, which is rested butt-down on the ground, barrel to the sky. To this day, he will still rise at dawn on the first Saturday of May to sit in a damp, camouflage-covered lean-to called a maimai on the edge of a small expanse of water and wait for ducks to fly in. Each year the start of duck-hunting season is a call to don arms. Ducks are maimed and sometimes killed in a four-hour flurry of lopsided gunshot.

These ducks are embedded in Professor McMahan's philosophical reckonings. Each sentient animal, be it a waterfowl or a cattle beast, has a right to a quality of life where they can freely exhibit their natural behaviours, on par with humans. That's the argument delivered with volume by vegan activists. When Professor McMahan talks of cattle farming, he sees conditions of 'considerable suffering', which is to say intensive commercial farming and specifically vast feedlots.

In Australia, 1.2 million cattle are 'finished' in feedlots. They spend a year grazing on pasture and the remainder of their life — between 70 and 300 days — in pens. In the US that number is closer to 14 million. Picture tens if not hundreds of thousands of cattle standing in dirt and dung-lined pens. They don't graze; they eat a medley of grains served to them in feed troughs lining the pens. Often, they are given antibiotics to stave off diseases like bovine respiratory syndrome, the most common cause of death in feedlot cattle. Professor McMahan considers this an environment of 'stress and misery'. So, he says, it's actually very simple, this decision we have to eat or farm cattle for beef: we can indulge in a diet that causes lots of suffering or have a diet that does not involve the infliction of great suffering.

I'm thinking about the calves we lost in the early days. There is no doubt in my mind that they suffered in the moments before they died. But I'm also thinking about all the other contented calves chewing their cud, eating grass, roaming across my paddocks, their presence and actions — dropping manure and urine — allowing us to decarbonise as quickly as possible by not using synthetic fertiliser or heavy cropping machinery, or buying in seeds in vast quantities to replant crops year after year.

Is their life full of suffering? I think not. Is their life without purpose? No. But this is a vision of animal welfare, not animal rights. The former, for more than a hundred years, has centred on a stewardship relationship between humans and animals.[3] I owe the animals good quality feed, grazing without danger or fear of injury, calm and safe handling, quick treatment to avoid unnecessary pain, and, when necessary, humane euthanasia: a good death.[4]

What constitutes a good death has been on my mind recently, because one point in McMahan's argument rings true: we kill a cattle beast for beef about one tenth of their way through their natural lifespan. But an increasing number of farmers don't think about the ethics of this at all. Researchers suggest that the intensification of modern farming practices — the pushing of cattle into feedlots, for example — has led to an 'instituted distance between livestock and farmers'.[5] The animal is an object that becomes a commodity. To think beyond that is to shoulder a moral burden many aren't equipped to work through. We kill cattle before their life is done. Why? I tell you it's to keep the ecosystem we've created in balance. But mostly, we do it for our own pleasure. A life cut short for our financial gain and culinary and nutritional satisfaction.

In the days after I miscarried my first child, before I moved to New Zealand, my face was gaunt, my pallor pale. I could feel my body waning. I was back at work the day of my visit to the emergency room. I wasn't in pain; I was depleted. And I craved one thing above all else: red meat. There is a nutritional reason for that. I was severely lacking in iron. I had lost a lot of blood. But there was a less tangible reason, too. It brought me a pleasure that for a moment squashed sadness. To eat is often to satisfy a craving. We luxuriate in flavours that pop and sparkle over our tongue, and we feel a rush of warmth as the food settles in our stomachs, making us replete. My pleasure was at the expense of a cattle beast's pleasure. My meal cut short that animal's ability to derive its own satisfaction from a meal of grass and hay.

In the first years on the farm, I flourished with the physical labour, the cold air; it made me strong again. It was a worthy endeavour. But there is a cost. Professor McMahan tells me, 'We'll look back on what we have done with horror and incredulity that we could have been so barbaric.'

And yet, horror is not what I see around me. But then again, I didn't follow the live export heifers to the feedlot or onto the boat. I don't work the line at the meatworks for which our animals are destined. I didn't go looking for the barbarism. I stay here, focused on what I can control and nurture within the boundaries of the farm. Is that akin to sticking my head in the sand?

There's a body of research that psychologists specialising in behaviour change are drawn to. A theory called cognitive dissonance, when one's actions do not marry with one's beliefs. Take, for example, wearing a seatbelt while travelling in a vehicle. We know, after decades of government-funded advertising campaigns, that wearing a seatbelt is not just legally required but likely to save your life if the vehicle crashes. It's a risk mitigation, and we can know this as a fact, yet we still on occasion choose not to wear a seatbelt because we tell ourselves 'It'll never happen to me' or 'I'm just popping down to the shop'. Excuses, excuses.

Australian psychologist turned politician Emma Hurst tried to explain to me that my cognitive dissonance about farming cattle and eating red meat might be tied to my nostalgia for a family farm long gone, and a pining for a romantic vision of the idyllic.

My farm is better, I keep saying. *My way isn't harmful.*

Hurst is a sitting member of the legislative council of New South Wales representing the Animal Liberation Party. She was elected in 2018 on a vegan-first platform and is in state parliament to represent the animals — farmed, domesticated, wild, the lot.

The New South Wales state parliament is located on Macquarie Street in Sydney's central business district. A colonial-era building once used as a hospital, it's surrounded by a wrought-iron fence; the fustiness of the ground-floor decor is reminiscent of my grandparents' home, built in 1886, on the family farm. The wood wall panelling of both the Parliament House and my family's farmhouse speaks of a certain age, when the creaking wheels of a five o'clock drinks cart was audible above the clang of the wind-up grandfather clock.

There are aesthetic and historic legal ties that bind New South Wales and New Zealand; the former governed the latter between 1840 and 1841. This was the time when the Treaty of Waitangi was signed and before land wars were fought in New Zealand and the violence of the colonial project truly descended.

As I waited for Emma Hurst in this panelled lobby, I thought about my grandparents' home and recalled the drinks trolley, not the cattle. Were there cattle there or it is a vision I've conjured? This is the challenge that confronts practitioners of behavioural change: how does one let go of the cultural, historical, philosophical baggage? 'Remove the barrier,' says Hurst. But don't force something new down people's throats, because that never works. Aggression never works.

In her parliament office with the ticking parliament clock above us, which reminds occupants of their obligation, we start talking about lion bones. That's her version of my seatbelt analogy. Hurst worked on a campaign to stop the sale of lion bones for medicinal use, which would, in turn, halt or at least slow the lion farming trade. Months later, Hurst sends me an Animal Liberation report that detailed a behaviour-change campaign targeting social media users who love eating chicken. The aim, ultimately, was to convince Australians to stop eating broiler chicken and to give a pledge to Animal Liberation to support their work. But the most effective campaign, in terms of engagement, was one that allowed consumers to meet the group halfway. To reduce consumption, and to educate themselves on industry practices. It's similar to the flexitarian-diet campaign.

Reduce.

Reduce your meat consumption, reduce your farm inputs, reduce your waste. But how does 'reduce' permit the killing?

By the logic of Professor McMahan, if one accepts that cattle, or any other animal we eat, are sentient beings with the ability to not only feel pain but live a life that contains emotional highs and lows — joy, playfulness, sadness, camaraderie — and that sentient animals have rights, then it is wrong to take an animal's life for the simple argument that it's food. Step down the sentient ladder and you're on safer ground; protein sources like mussels, clams, and oysters become an ethical and nutritionally satisfying meal because molluscs don't have capacity to think. The sentient-being argument is clean and hard to counter. Yet it leaves little room for ethical concerns about the human–animal relationship in combating global warming. Does this pressing issue give us room to deviate from the moral certainty of 'meat is morally wrong'?

I asked Professor McMahan how he envisages an end to cattle farming should we all stop eating meat. Stop the breeding, he said. It's a long play, not swift action. Just let the current global herd of 1.4 billion grow old and die. Some will live for 20 or more years continuing to emit methane, and then the feedlots and farmlands will be empty of cattle.

Temple Grandin, an animal behaviour specialist of renown, who works with ranchers in the United States to create animal-centric farming systems, says that were it not for our collective desire to eat meat, cattle would not exist. If they didn't exist, they would never get to experience the joy and contentment animal rights activists advocate. But with the end of the meat trade comes the end of cattle, unless some are left to roam in the wilderness — and what then is our moral responsibility? Without an apex predator, do they become a pest like so many other introduced species the world over? Is it then our responsibility to control the pest to keep the ecosystem in balance — to be the apex predator?

There are stories shared among hunters in New Zealand of bush cattle that can be found in the deep forests, hours' trek from any domesticated land. In the 1930s, crippled by the economic fallout of the Great Depression, some settlers simply walked off their newly broken farms as the return on exports, especially wool, plummeted by up to 60 per cent. They left behind cattle, which took to the bush, bred, flourished, and turned mean. These big beasts are now one of the most dangerous animals hunted in New Zealand, but if a hunter drops one, the meat, apparently, tastes like native pepper tree.

To hunt is an adrenaline sport for some; for others, it is an act of conservation, in aid of protecting the fragile plant and bird species that exist in the native bush, and which have, for more than a hundred years, been ravished by introduced pests: red deer brought over by British settlers for the purpose of 'game' hunting; weasels, stoats, and possums, thar, pigs, and wild cattle all do vast damage to diverse habitats. One hunter, Jannine Ricketts, told me that while hunting provides her family with food, it is also 'an important role as a kaitiaki to look after that ngahere environment', which is to say that to hunt is to act as a guardian for the forest. To protect something complex and precious.

The global push to create more farmland has fast-tracked deforestation and caused devastating habitat and biodiversity loss. Biodiversity is being destroyed by humans at a rate now unprecedented in history.[6] One counter, which can be heard with volume in Britain especially, is that the answer to our ecological woes and dramatic loss of biodiversity is to rewild vast tracts of land. But that presents another ecological conundrum: who or what will manage these new forests to ensure balance? Do we simply lock up land and expect nature to do its thing? We humans ripped the natural order apart long ago, and transported animals and plants around the world under the guise of societal and economic development. History shows that if you lock up land once cleared in New Zealand, what you get, within a decade, is a cacophony of colonial imports squashing native life: a forest of wilding pine trees home to possums and deer. Domesticated land, because of our collective impulse to dominate it for so long, still needs stewardship, guardians: people with the vision to act as kaitiaki, to work with nature's rhythms and not against them.

Jannine, who is a winemaker as well as a hunter, is part of a biodiversity group that has embarked on a replanting program in a block they hunt. Its members are putting thousands of native plants in the soil and holding back the introduced gorse through manual labour. The recreation of nature is as large a task as the destruction of it. It's not enough to lock the gate and turn away.

These acts of planting and weeding are small gestures, but there is a rebalancing that reveals itself in the presence of tui birds and monarch butterflies; both have settled in around our farmhouse of late, captured by the food corridor that has appeared as trees and shrubs have been planted along waterways now stretching east and then south, towards unmilled native and pine forests. These birds and butterflies now fly habitat routes that run like arteries around farmland.

And I watch.

The creation and maintenance of what I think of as 'new nature' is an act of supremacy that falls out of step with the animal ethicist's equity argument. Stewardship, as much as farming for productivity, is a role of domination but for different ends; one value system privileges the environment, the other profit. As I reckon with the act of killing an animal for food, I also grapple with a history of supremacy, and it cuts me off from Professor McMahan's elegant philosophical theories.

When I spoke with McMahan about the practical ethics of eating meat, and, in turn, the farming of animals, we talked a little about the notion of reparations to right historical injustices, and the right indigenous peoples have to claim they have been disadvantaged by the colonial project. I was thinking about the conversation I had with James, the chairman of the Māori trust that runs a large dairy farm near me. He was adamant that the land they have, which has been hard won, must remain as productive as possible, because the benefits of profit have the potential to lift a generation up through scholarships, traineeships, and rehabilitation programs. And the most productive version of farming on his land is dairy, with a herd that numbers in the thousands, in an intensive system similar to those McMahan has associated with barbarism.

In thinking about historical reparations, McMahan sees the process as a complicated philosophical problem about justice and the distribution of benefits and burdens. Had the Crown not engaged in the colonial project in New Zealand, Australia, Canada, the United States, and elsewhere, different people would have married, different children would have been born, and so, he says, 'We have to take into consideration the fact that the alleged victims of the injustice who exist now wouldn't have existed had it not been for the injustice.'

I had asked Professor McMahan if the financial benefit for a Māori tribe long disadvantaged by the loss of land and wealth that came from the farming of animals could outweigh the harm done to the animals in the act of slaughter. It was a crude equation: animal lives for the wellbeing of an extended family and community. Death for life and opportunity. Is there a greater good in passing on the burden to another group? McMahan says no. There is little justification for passing the highest burden of death onto an animal for the community benefit of living well, or better than the generation before.

Arguments of ethics often fall on conclusions that bang up against one's value system, and this is where my heart departs from McMahan's theory. The injustices of colonisation cannot be rubbed clean by philosophical reckonings. There are too many damning statistics in morbidity, housing, food insecurity, and poverty that show a gross disparity between indigenous peoples and settlers who've generated wealth and advanced most swiftly through private ownership of land and the subsequent guiding of nationhood.

Pulitzer Prize–winning American author Isabel Wilkerson says that contemporary society has formed into a caste system, a concept that runs through colonial societies split by race and class. She considers caste the underlying skeleton over which lies a skin of race, and racism. One is born to a caste. It is, Wilkerson writes in her book *Caste: The Origins of our Discontent*:

> an artificial construction, a fixed and embedded ranking of
> human value that sets the presumed supremacy of one group
> against the presumed inferiority of other groups on the basis

of ancestry and often immutable traits, traits that would be neutral in the abstract but are ascribed life-and-death meaning in a hierarchy favoring the dominant caste whose forebears designed it'.[7]

What of these immutable traits? Is an ingrained commitment to collectivism and a connection to land — a way of knowing and never owning — one such immutable trait?

When I spoke with Māori farmers Dominic and Wiari, they told me about the trouble their old people had securing bank loans when the government demanded the land be developed as productive farmland. The banks refused them because the land has never been owned by one individual, a single person liable for debt. There has been one set of rules for those who live bound to land and a web of familial connections; another set for those who stand alone above everyone else, as Wilkerson points out. Financial lending is just one institutional system propped up by the assumption of supremacy — there are many others.

I didn't ask Professor McMahan if he would have reached the same conclusion if the end goal of reparations was not profit and opportunity for a disadvantaged people but rather a flourishing ecosystem. The death of a cattle beast before its natural life for an ecosystem thriving would still result in 'No'. If you accept that sentient beings have rights, and those rights include the ability to live a full and content life free from suffering, then it is morally wrong to kill and eat that animal, no matter the end goal.

And so I assume, or rather retain, my position of supremacy. Over the animals, but in service to the land. Wiari taught me a saying that guides his thinking about the future: *Manaaki whenua, manaaki tangata, haere whakamua.* Care for the land, care for the people, go forward.

So, I farm to kill.

—

There's no justice in killing an animal. That's what Scottish Louise Gray, journalist and author of *The Ethical Carnivore*, says. For an entire year, Gray consumed meat only if she had killed and prepared it herself. It may sound like a grim project, perhaps even pointless to some. But it was born from a desire to fully understand why the eating of meat had become an act that had to be weighed on an ethical or moral scale.

Gray is now long past mulling over the philosophical and ethical right and wrong of it all. She's resolved that there is no justice in going after an animal with a man-made weapon, as one does when stalking a deer in the hills of the Scottish Highlands with a gun, to kill it for food. But if you're going to eat meat, you must accept that the kill is part of the process. And acknowledge that at some point a sacrifice, by the animal and the person[8] doing the killing, is being made.[9]

There's a historical hierarchy that runs through the food system, be it in New Zealand, Australia, the United States, or the United Kingdom. And in simple terms it's this: landowners and farmers have long outsourced the killing of their animals to centralised processors that rely on a workforce to do a job few want.[10] The work is laborious and taxing on the body and the mind, and it's fallen on men, typically, from immigrant or indigenous communities, who have been drawn from their own land to urban centres with the promise of steady work. Yet that steady work has historically condemned them to live at the fringe of society, according to writer Noilie Vialles. She argues that nineteenth-century European attitudes towards death demanded abattoirs be pushed to the boundary of modern urban life; Britain took the modernism further, and established farming and processing environments on the other side of the world. Vialles writes that what was once a bloody spectacle, and before that, a ceremony to be carried out among community, is now rendered invisible, and the consequence for some workers is disassociation from the act and the animal. This disconnect, unsurprisingly, has been linked to cases of post-traumatic stress disorder among workers.[11]

The food system is entwined with the history of colonisation, land rights, urban drift, disadvantage, and supremacy. And my place within this system, as it has been since my ancestors arrived in New Zealand in 1881, is defined by my ability to choose. To choose whether to kill or not, and to choose what to eat each night for my emotional and physical health. I've never known the moral burden of taking a life. Yet if I'm going to outsource the killing of animals then I need to know what is sacrificed — in the animal and in me — during the process.

If you want to learn how to kill a chicken, google it. There are hundreds of homesteading blogs and farm-focused Facebook community pages that provide step-by-step guidance for 'humane' killing. There are also regulatory guidelines outlining the steps that need to be taken to avoid unnecessary suffering, and to comply with animal welfare laws. I learn on a small-landholders blog that one option for dispatching a chicken is to use a broomstick. Not as a bludgeon, but as a sort of suffocation tool that works to compress the neck, and then, with a swift tug of the legs, dislocate the head from the rest of the body.

Another option is to repurpose a five-litre plastic water bottle and turn it into a kill funnel, for want of a better word. By removing the lid and part of the bottleneck, you create a hole for the chicken's head; by slicing off the base you create an entry point for a chicken to drop, head first, into the container. Attach that funnel to a tree or something sturdy and you've got yourself a contraption where, according to the blog, you can dispatch a chicken with one quick flick of a knife. I know all this because I've been reading up, preparing to kill our layer hens that have been living in my back garden, eating food scraps from my kitchen, for the past two years.

I catch a hen, which has long stopped laying eggs, one morning when the temperature has barely pushed above zero degrees. Five of my fingers are numb and white at the tips, which seems like an ominous sign. In one gloved hand, I hold its head, so it won't peck at me. The other is grasped around its body. It goes into the plastic container smoothly. Head first, legs up. But unlike the diagram, the head doesn't drop straight down; instead, the hen is tucked up tight into the container.

The adage 'if you stick your neck out …' is on repeat in my head. A strange motto for the moment, since it's not in the chicken's best interest to take a risk right now. Minutes pass and the head finally drops down, and I grab it and run my knife. It's done, but not easily. Next comes a process perhaps worse than the initial act. Bleeding out, feet and head off smoothly with a cleaver. Skin sliced open at the breast, and the chicken is de-gloved while warm. Out come the insides, and what's left is a tiny chicken of yellowish colour. It looks little like the plump, white store-bought versions.

Two days later, after the bird has rested in my fridge on a plate under clear wrap, I make stock from the meat and carcass. I put the carcass into the pot along with onions, celery, carrot, peppercorns, salt, and thyme. The result is a pungent, slightly gamey broth that is both delicious and sour in my mouth. The latter is my brain overriding my tastebuds. I should be luxuriating in the experience of eating food that I know with absolute certainty has come from a free-range animal free from antibiotics, hormones, and chemicals. But the metallic funk that lodged in my nostrils at the time of death has returned.

I only killed one. The other four hens clucked and scratched in the yard for another two days until Pat gave up waiting for me to finish the job I promised to do. A sharp tomahawk, a wooden block;

four headless chickens. Swift and efficient. I buried them, intact, in the garden.

I recall Louise Gray saying that she derived no pleasure from her hunts but that feeding the meat to friends and family was deeply satisfying. 'It's a huge moment to source food for those that you love,' she says. Yet it's not a moment for everyone; Gray and I know this. My hesitancy is cruelty. My lack of skill makes me unfit to dispatch an animal. The work done by abattoir workers is a skill that should be respected by community, yet it's barely recognised, because we've collectively forgotten how to acknowledge and appreciate a good death. If we were to pull abattoirs back from the shadows, confront the process, and honour the work, perhaps we'd collectively elevate meat to its appropriate position: a rare food that requires sacrifice. Something fit for eating with friends or family occasionally, or during ceremony.

Eating meat nightly is not an unfettered right; it's a habit made possible because death is banished to the outskirts and most of us choose not to look for it. We prefer to think that meat comes wrapped in plastic.

Towards the end of winter of the first year of the COVID-19 pandemic, the heifers we accidentally acquired as calves, that I mistook for bulls, had grown fat and muscular. They were the first two cattle we sent off the farm to the meat works to become food for others. In the report that came back from the abattoir, we learned the heifers had been classed as P2. A good amount of fat, not too much muscle development. Right for the local market — in a country that relies on export dollars, we decide to provide meat for New Zealanders instead.

Yet in that month, the workers breaking down, butchering, and packing our meat were among the most vulnerable in the agriculture supply chain. From the United States to Australia, the virus thrived amid the assembly line, the poor ventilation, the cold temperature, and the overcrowded lunch areas of meat processing plants. In the rush to recreate something nourishing within the boundary of the farm, I'd lost sight of the community outside.

The season I started digging for worms was also the first months of my daughter's life, and I was eating more food than I have ever in my life. I was constantly hungry, a deep low rumble of want. On the weekends, I'd set a slow-cook brisket on for dinner. Searing the hunk first, I'd slather it with a pureed mix of chargrilled onion, garlic, red pepper flavoured with toasted coriander and cumin seeds, salt and pepper, chilies, sugar, a dash of soy sauce, and apple cider vinegar. Next, I'd add water to the bottom of the pan, submerging half the meat. I'd cover the oven dish to let the steam infuse the meat, and then roast it for hours, until the meat fell into slithers.

We eat morsels between homemade tortillas; I stain the flour pink with beetroot for my little girl, who delights in scooping little handfuls into her mouth, succulent meaty juices running down her chin. Meat is a feature of our diet, but it's a slow luxury now, not a rushed daily dish. Cook, writer, and broadcaster Rachel Khoo told me she thinks about meat as a condiment, something that adds a layer of flavour atop a dish. It's a sentiment I've grown to appreciate as we scale back, slow down, and consider the sacrifice made for this meal. To eat a little will change the food landscape in time, if enough of us do it. We are beef farmers who want people to eat less beef.

We keep sending cattle to the works, even though a petition is now circulating. Started by an anonymous leader, it is calling for the government to shut down New Zealand's meat-processing industry to prioritise workers' wellbeing and safety over the supply chain. More than 2,500 signatures are collected, but the abattoirs stay open, deemed essential by the sector and the government. For those working inside the plants, the industry motivation for continuing

— export profit — seems far removed from their priority: family safety. Signatories start leaving comments on the petition:[12]

I don't want to risk spreading the virus to people and we are not essential to NZ. 90 per cent of the meat we process go [sic] off shore. (Richard Green)

Over 90 per cent of product is being fucking exported overseas, fuck the foreigners look after our own first. (Ethan Ellis)

Ricki-Lee Corcoran writes simply: *Whanau first.*

Family first.

Locals first.

The land inside the gate is flourishing; we are getting closer each day to a farm that gives back to the environment, rather than constantly takes. But as soon as we step outside the gate, the gains we've made by turning our backs on conventional farming are squashed by a more pressing issue: the food system is broken. As food tech entrepreneur Josh Tetrick says: it isn't fair, it isn't equitable, and we need to fix it.

Is it a farmer's role to care for their community as well as the land? To produce food to feed locals? To ensure they have enough in a time of need? In the moment I say yes, loudly, I'm labelled an activist — a leftie with an agenda, because I'm prioritising the local food chain over the lucrative export trade.

How did 'local first' become such a profound statement of protest?

8.

Global and local

My father likes to chat and impart knowledge. He settles into one of the wooden chairs around the outdoor table on our veranda and leans forward, propping himself upright on his elbows.

'I don't know if this is relevant or not …'

A harrier hawk is dive-bombing a gosling in the distance behind his head.

'You know the woman up the road? She's fattening her cattle on fodder beet. She just sent some to the works, got something like $6 a kilo for them. More than $2,100 each.'

We don't grow fodder beet for our cattle; we don't crop anything for our cattle to eat. It's an environmental burden we're not willing to drop onto the land. Yet the insinuation is that I should just because someone else does. Push the land to its capacity, that's the tradition of New Zealand farming. That's local farming.

Fodder beet is used a lot in this part of the country to feed dairy cows during winter. Herds are given access to a strip of beet, and then moved off or given something else, like hay or silage, to chew on in between. The enormous tubers are high in sugars, and if you get the feed routine wrong, the cattle can die of acidosis. Too much sugar, basically. Get it right, and the cattle stack on weight over winter, a time when the growth is limited. But then the paddocks are left exposed, and the soil freezes in the early spring chill before it's replanted with new grass. It's a cycle of inputs and labour.

The fields in the region we farm at a certain time of the year glow vivid green with feed crops; in late spring, the countryside is spotted with burnt brown flags signalling the charred remains of a grass paddock about to be planted. This is the pattern that's passed on as knowledge by sales reps, industry bodies, and farmers themselves.

Don't forget the slug bait, my dad says. 'I didn't bother one year, and the buggers ate half of it before the cattle got there.'

I have an Instagram account for our farm. My Australian cousin once asked what the point of it was, insinuating that everyone must have an agenda on social media. It's a marketing tool, I told her, in case we one day have something to sell. But mostly it's worked as a visual diary since we arrived. My grandfather used to keep a written farm diary. Detailed notes on weather, stock movements, weights, cropping rotations, soil quality, sales, fencing requirements, burn-offs, staff comings and goings were all recorded, every day. A stack of notebooks was found alongside black-and-white photos and glass plate negatives in the attic, long after my grandparents had left the farm to retire. An attic of knowledge and memories left to gather dust and then be thrown out. Discarded and relegated to the status of something we once had.

One day, a woman I had met while at university sent me a message on Instagram:

So interested in following the journey of people choosing the farm life who aren't farm owners. I feel like there is so much focus on farmers as only pursuing it because of the growth in capital land value. So keen to hear more about your proposition.

My proposition is to choose to farm. To *learn* to farm, without much of a guiding hand or the weight of familial obligation.

How Pat and I accumulate knowledge is a mix of inquiry and disbelief. So much of what we've been told doesn't seem quite right; but it's merely part of a knowledge tradition that we sit outside of — one focused on science funded by the agriculture industry to maximise yield. I've read that some farmers privately discuss the difference between a good farmer and a successful farmer.[1] The latter push the limits of nature as far as they bend in the name of profit and productivity, hoping all the while the system doesn't crash. The former work within the limits nature has set out, which comes by learning from the 'old fellas behind ya' as Bill Coogan — the hill country free-range grazier I met through my uncle — describes it. 'And learning from your mistakes,' Jennifer, his wife, confirms. This method assumes an unbroken lineage where family are farmers generation after generation; it mirrors in a small way how indigenous knowledge is transferred. Both have for a long time been relegated to the tier below scientific research. Knowledge gained from being in one place for a long time is somehow lesser.

Dr Laura Pereria, a South African researcher based in Britain, writes that the transformation from unsustainable environmentally damaging food systems to sustainable systems can start at the point where we collectively recognise the innovative potential of diverse local and indigenous knowledge systems.[2]

To reframe the food system as something sustainable and local is also to rearrange the knowledge hierarchy: to listen to voices that have been softly whispering, not loudly boasting. But what does local even mean? Like every other question I've had in the past two years, the answer depends on who you ask.

I've been experimenting with a recipe for rewena, a leavened bread made from a potato starter. It's an old recipe, made new in a book by New York–trained New Zealand chef and author Monique Fiso. She tells me to cook urenika — a purple Māori potato — and then follow the instructions for the starter. And wait. A week at least. I use white-flesh agria potatoes instead.

Wheat flour was one of the first things to disappear from the supermarket shelves when COVID-19 shuttered shops and borders, but only temporarily. We were assured that the supply chains were solid; the logistics and packers just needed to catch up. Both New Zealand and Australia are net-positive food producing countries, but if the import and export lines cease, New Zealanders especially will notice the shortfall. The country imports a third of its foodstuff, and within that third are key staples and beloved luxuries: wheat, refined rice and legumes from Australia, bananas from Central America, sugars, pasta, carbohydrate-heavy processed foods, and, ironically, European-style cheeses that are often sold at a much lower price point than local dairy. If we didn't import, our collective diet would be vastly more restricted. But perhaps it should be.

When I worked at BuzzFeed, a journalist colleague named Amy McQuire embarked on a pilgrimage of sorts to Uluru with her young daughter. There she found the kitchen — a waved rock where women worked thousands of years ago. Within the ancient formations were indentations, large grooves hollowed out into grinding circles where, McQuire realised, seeds would have been ground into flours to make bread. A local bread, using local ingredients, bound by local knowledge. 'You can sort of feel them,' recalls McQuire, who is a Darumbal and South Sea Islander woman. 'The old people are still there.' And so, too, is the food knowledge of that part of the country. 'It's just been kept away from us for so long,' she says.[3]

As the fallout from COVID-19 continued to ripple through the community, calls of 'local first' grew louder and more sustained.

Support local business, support local jobs, support the economy, eat local food. Food sovereignty, not just security.

But what does that mean in the era of a globalised food economy — to have sovereignty over food?

For Fiso, whose family heritage is Samoan, Māori, and European, to eat local requires decolonising the palate. She describes a hummus-like porridge made from corn left to ferment in fresh water for six weeks. It's an acquired taste, due to its unholy fragrance; it's rotten corn. Yet the culinary journey back to Māori foods is an act of sovereignty as much as a gesture of gratitude. In her book *Hia Kai*, Fiso elaborates that where there is food, there is a spirit of hospitality that runs deeper than the sharing of a meal. Food can bind people who are split by intangible divides: 'we build unity through the act of giving and sharing', she writes in a passage that gives thanks to ancestors and the gods who handed down the treasures that become food.[4]

My rewena bread is a funky, stodgy disaster — I should have followed the recipe. But my attempts to buy any variety of Māori potato, as the recipe required, were fruitless. They are a rare luxury, I discover. Old local foods, like Māori potatoes, exist cherished by family, not industry. They're grown in potted buckets filled with compost, not across thousands of hectares. A potato farmer explained to me that despite their superior flavour, popularity at organic farmers' markets, and high yield, they are not commercially viable. Some are prone to disease and become a loss leader. And, unlike varieties that have license fees attached, there's little interest in putting taewa — the generic term for Māori potato — through a lab process to strengthen the seed stock, because the financial return for the corporate entity funding the research is nil.

Taewa aren't owned by any one company; the varieties aren't patented. They exist as a trace of early trade and food resilience among Māori. The tubers are endemic to the Andean region of

South America and were traded into Aotearoa. Once in the soil here, the tawea proved more resilient and reliable than the kumara sweet potato, a staple in early Māori diets. To eat local is not to eat tawea, I discover.

Assuming 'local' equates to 'heritage' food is a common misconception, but we don't need to untether completely from the global food system. To do so is to erase the long tradition of trade and food innovation that existed well before European settlement, and risk more food insecurity among families barely getting by. Dr Lisa Te Morenga, a nutrition and Māori health researcher at Massey University, points out that Māori cuisine has always evolved: Polynesian seafarers traded for tubers, settlers introduced new carbohydrate and protein sources. 'Māori were growing successful market gardens at spots in Auckland with imported foods that were just easier to grow in our climate,' Te Morenga says.[5] To eat local is not to suspend time — it's to innovate for nutritional and cultural sustenance, something we need desperately.

Currently, only 60 per cent of New Zealand adults eat three-plus servings of vegetables a day;[6] the figure is comparable in Australia. The number of adults who follow the good-health dietary guidelines of five servings of vegetables and two of fruit is minuscule: around 8 per cent[7] of the population, as of 2018. Both Australia and New Zealand bear a malnutrition burden that presents as anaemia among teenage girls and women, obesity, and in New Zealand especially, concerning rates of non-communicable diseases such as diabetes.[8]

Two nations of abundance, populated by undernourished people.[9] Yet in both countries, the value of food has never been solely about sustenance. Its main value is found in its trade, and that's how the jobs and economic input continue, often at the expense of locals' health and cultural wellbeing.

Most people in the world rely on distribution mechanisms and import markets for food staples. Less than 30 per cent of the world's population can source their temperate or tropical cereal grains, rice, or pulses within a radius of 100 kilometres; most source their staples from producing regions more than 1,000 kilometres away.[10] This can be seen as a flaw in the global food system, but it's also proof of the success of the Green Revolution. Trade profit came from homogenising the global palate.

In a 2014 study, that trend towards a global taste emerged in a stark crop list. The researchers measured the median change in developing countries of crop to calorie contribution between 1969 to 2009 and found wheat to be the new major staple in all countries. Meanwhile, crops with long-held regional and cultural importance, such as endemic tubers, sweet potato, cassava, and yams, once staples of the Pacific and Australia, were disappearing from fields and diets. Yams, down 31 per cent. Coconuts, down 33 per cent. Sweet potato, down 38 per cent.[11]

When Monique Fiso suggests that to eat local is to decolonise the palate, she doesn't just mean we need to acquire a taste for rotten corn. The target is not just the dinner table and the kitchen but the field: we need these crops back. We need a market system where something other than commodities — wheat, corn, soy, sugar, beef, dairy — can flourish.

My gran wasn't much of a gardener. Two kitchen gardens sat behind the farmhouse, and the family's bedrooms looked out over the produce that grew each season, but the grower was my grandfather. It's difficult to grow food; I've planted enough seeds over the past few years to know that failure is the norm. I don't have the patience of my grandfather, who would treat those small plots as an extension of the farm, creating boxes of fertile soil that were planted, rested, nurtured, planted again.

My gran wasn't a gardener, but she was a great cook. She would talk to herself as she stood at the stovetop, scraping the sticky pan juices away from the centre so they bled into the added gravy at the side of the pan. In lieu of having others to talk to, she'd work out her problems over the stove. I take after my gran in a way, constantly whispering to food. My mum takes after my grandfather. A devoted gardener and nurturer. And she holds both the maternal and paternal knowledge that trickled down until the line was severed, when I left New Zealand.

Now with a timidity that belies the value of the information she has kept close, my mum is guiding me in subtle ways: scolding me gently for planting tomato seeds too soon in spring before the last frost; reminding me to feed my lemon tree and vegetable plants with the ash from the fireplace to replenish the potassium stores depleted in the volcanic soil; pointing to the hills where red-top plants are sprouting, saying the pH must be off; cajoling me to leave my parsley plants alone until they've grown higher than scraggly stumps. 'You'll have more herbs if you stop picking the new growth,' she says.

Frank, practical, simple — but not uncomplicated. It's knowledge to grow food around the house and on the farm. The very thing I came here to do, and yet I listened to everyone else before I heard her words. She would float tidbits as throwaway statements, because I didn't ask. Initially. The hierarchies embedded in the global neo-liberal knowledge system aren't geared to honour the words and wisdom of a 71-year-old woman.

She is a patient guide, and finally I am here to learn how to appreciate local; to grow and eat for my context. But her advice is specific to right here, where I stand and write. Would it have made sense if I'd asked while still living in our third-floor apartment in Sydney? No. Pat and I didn't garden; instead, we'd visit the Marrickville farmers' market to buy eggs and niche products from small-scale farmers, who set up shop so passionate foodies can enjoy the romantic, agrarian-infused shopping experience and pretend they're 'shopping local'.

The urban farmers' market has been positioned as the site of equity, the place where knowing foodies can buy direct from farmers and protest the chokehold supermarkets have on food distribution. And in a small way, those who sell and shop at farmers' markets *have* succeeded in spotlighting traditional food and ecological knowledge.[12] But it hasn't done much to destabilise the global food regime, nor put more money into the pockets of farmers who are providing the quality ingredients because, more often than not, they don't or can't afford to pay themselves for the hours spent selling at the market. It's a site of rural nostalgia to wrap around the goods on sale.[13] Yet it's not much more than a convincing image.

In a small survey of 263 respondents conducted by Open Farms NZ and Our Land and Water, 29 per cent noted that the best single action they can take to promote sustainability is to buy direct from farms. But a web of small farm-to-foodie businesses will not remake the food system into something more resilient, sustainable, and sovereign. The global food regime will not be displaced by the romance of smallholder farming and farmers' markets. The knowledge base for building something new is not found in the nostalgic past.

The difference between food sovereignty and food security is in the relative importance of the market, writes Hannah Wittman.[14] The concept of sovereignty asserts we all have the right to produce the food we want when we want, and to do so in a sustainable system; to define the system we see as best. In a colonial country, food sovereignty is yet more powerful when the definition is extended to producing food from land one *belongs* to, not just owns.[15] Food security, on the other hand, presumes access to food is a problem to be solved, and that the solution can be found in trade. Food is a commodity that can be sourced to ensure a nation is food secure. This is the reality New Zealand and Australia rely on as exporting nations, and the route the United Kingdom pursued to ensure a steady supply of cheap foodstuffs.

Professor of food policy Tim Lang, one of the authors of the influential EAT-Lancet Commission report on the future of food in the age of warming, reflected on the United Kingdom's food dependency in the report 'A Food Brexit: time to get real'. He and his co-authors pinpoint the moment the UK relinquished its food sovereignty to 1846, with the passing of the Repeal of the Corn Laws, which gave Parliament avenue to buy food from whichever country offered it most cheaply — a by-product of the UK's imperialist outlook, which included having access to foodstuffs from its rich colonial network.

For a short while, post–World War II, the government thought growing more food locally was a good idea, but for the most part UK governments have favoured what are often termed 'cheap food policies', because they have supposed that contributes to keeping UK labour costs low, and thereby helping UK firms compete in export markets.[16]

To play in the export market is also to play nice at the World Trade Organisation (WTO), to broker and agree to multilateral trade agreements and to avoid trade wars, which Australia and New Zealand do well, mostly. The implication of including agriculture and food production in the WTO, however, is to enhance the divide in opportunity between the global South and North. The South gets stuck in commodity patterns to ensure countries like the UK have food. This disparity is where the rumbling is, and the stand for sovereignty starts. As much in global South countries as in rural areas of the UK, where post-Brexit, farmers, food activists, health experts, and environmentalists have formed an uneasy coalition to try to protect the country from what they see as an onslaught of unsavoury food made and produced in conditions not up to the UK's high animal welfare and environmental standards — that includes New Zealand and Australia, both recent signatories to free trade deals that will see commodities enter the UK tariff-free.

In the wake of the 2021 Australia–UK trade deal, Red Tractor — a farm certification body — started issuing social media announcements that Australian produce fell far short of the animal welfare and environmental standards British farmers follow. *Hormone-fed beef is legal in Australia*, it shouted. *Hot branding, sow stalls, paraquat weed killer, all legal!*[17] Red Tractor was established in the aftermath of the foot and mouth disease outbreak in 2001 that resulted in the culling of 6 million pigs, cattle, and sheep across Britain. The outbreak ruptured consumer trust in British food products, which took years to rebuild. This social media opposition is essentially a campaign for food sovereignty, one that implores Brits to seek and appreciate the cultural value buried deep within their farming system. The bullseye for this campaign is feedlot beef and factory-farmed poultry from America and Australia, and lamb and dairy from New Zealand.

In April 2020, when the United Kingdom was swamped by a wave of COVID-19 illness and Brexit negotiation anxiety, a rural journalist sent out a tweet that exposed a festering wound. It read:

No British lamb on the shelves of @LidlGB or @sainsburys Shrewsbury stores this evening. Instead full of NZ lamb when Easter is this weekend. I'm absolutely disgusted! Why are you not supporting our farmers who right now need our retailers to back British farming? #FeedTheNation.[18]

I read it and initially thought, 'Great, on top of everything else, now UK farmers hate us too?' It can be read as a desire for protectionism, but what it is, in fact, is a cry to value a local food system in jeopardy. When Pat and I chose to sell our beef to customers in New Zealand, to locals, over the more lucrative export market, we could have written the same tweet. Buy local, not imported, or else we'll fail.

On the morning I realised our fledging beef box business wasn't going to work because the mobile abattoir had pulled out, I sent out more than 50 emails with the subject line: MARCH MEAT BOX REFUND. It read:

> It's with much regret that I write to inform you that we've issued a full refund on your Slow Stream Farm Seasonal Meat Box Order. After a number of Covid-related delays, we've today been informed that our processing partner — the mobile abattoir — is unable to travel to complete our job because they do not have access to the expert staff required to proccss our animals on the farm. This is an unexpected development. When we released our Meat Box product for sale we were confident that our supply chain was strong. But the mobile abattoir, like us, is a new business and they too are working through some kinks in their system ...

'Kinks in the system' was the reason I gave to supportive customers, who flooded back with responses of support. 'Please just keep the money,' one wrote. 'We believe in what you're doing'. I issued a refund nonetheless, the nausea of failure rising like hot bile in the back of my throat.

What was it we were trying to do? What was it they thought we were doing?

The idea for our farm-to-plate meat-box business was simple: provide affordable, quality meat to locals, meat that was farmed well in a regenerative system. It was to be, anecdotally, 'stress-free', because the kill was to be done on-farm. And people wanted it. Supported us. Told us to keep going because they saw flaws in the current system. We received emails from all over the country from people wanting our meat, people wanting to visit the farm.

Why?

That morning I fixated on all the hours I'd spent developing our little meat-box business only for it to fail. Boxes containing wool wrapping and cool bags sat unopened in my office. Totems to an idea, not an action. But failure is just the process of learning; I was reminded of that by my baby daughter, who was learning to walk, starting with a wild-legged wobbly stance of pure grit.

'Don't give up,' the email said.

Pat told me something different: 'It's okay to stop'. I was running through alternatives: sending animals to an abattoir to be slaughtered and returned as carcasses to our butcher. But each step was replicating the status quo. We were trying to create something new; I'd just gone too fast.

The boxes were put to one side. Keep going, they said. We want to be part of your community, they said. We'd spent so many months shut up behind a gate making changes that made sense for the land that it hadn't occurred to us that people were looking in through the fence wires. Noticing. Witnessing. Wondering why our farm was starting to look different to the one next door.

It's so easy to believe that a great divide exists between those who farm and those who eat. But if you open the gate, the curious will come.

Robert Pekin, the founder of the Brisbane food enterprise Food Connect, discovered this decades before me. He told me people are drawn to his way of operating — the cutting out of the middlemen: the wholesalers, processors, distributors, and large retailers — because they want to be 'in community with farmers'. That drive, to understand not just where their food comes from but who's produced it, what their challenges are, what keeps them up at night and causes them to celebrate, is the binding ethos of Pekin's farm-to-food-hub system.

Boxes of different sizes can be purchased from Food Connect, and each week a list of items is distributed alongside the name of the farmer providing the crop and the location from where it comes. When Pekin established Food Connect, a driving motivation was to pay farmers a fair price for their produce. Forty to 50 per cent of the retail price goes back to the farmer. The average return in Australia is 14 cents on the dollar; in the rest of the West, it's around 10 cents.

If you cut out the proverbial middlemen in the hourglass system, the food remains at roughly the same price for shoppers, but more money goes back into the pocket of farmers, and that has the potential to disrupt the entire system. Traditionally, producers have little to no negotiating power because there's only one or two buyers willing to take their product. And the product that is taken is graded by a retail system geared toward perfect produce: no blemishes, no misshapes, nothing too ripe. Food is dumped and dropped because it does not meet an aesthetic standard.

Robert operates differently; he'll take all of it. Paying a flat rate for the crop with a dollar bump for the best of the lot. This gives the farmer the chance to forecast return, plan for improvements on the farm, and pay themselves a living wage. And Robert is charged with persuading the consumers to eat real food, food that doesn't look spit-polished.

It's a system that draws learnings from the ethos of community supported agriculture (CSA) — a style of food production that allows consumers to buy into a single farm, giving the farmer capital to produce food for their subscribers. Each person or household receives a box of goodies for their financial contribution. Some trace the modern CSA movement to 1970s Japan, but the system has roots in the southern states of the United States among Black communities who, in the 1960s especially, relied on the farm-to-plate model for sustenance.

Robert ran a CSA in partnership with two others in Tasmania for a few years, but he found the system at times stifling. He wanted to change the whole foodscape, to work with a range of farmers struggling to make a profit under rigid retailer contracts and for consumers who 'hated Coles and Woolies'. Before he retired from his CSA, he sat with an older woman who'd followed his journey, and she counselled him to slow down. 'You're so bullish about changing the world,' Robert recalls her saying, 'that no one has a chance to keep up with you.'

Now Robert is back thinking about how to change the world, and he's planning to open food hubs in main centres around Australia. He reckons the people most able to bring others along on a change journey are those who fail. Fail often and expose their flaws.

'I think the world would just be a lot better place if we got over this idea that we have to be perfect,' he says. There is empowerment deep within a shared and humble story of failure.

We were talking about failed crops when Robert shared this insight. He has, over the years, had to convince organic farmers he works with to sell him produce they see as a failure. Four acres of corn not up to size, but beautifully juicy and sweet, was to be buried intact by the farmer, because he was too embarrassed to sell it under his farm name to his usual organic wholesaler. Robert took the lot and explained to his customers that the corn was stunted because nature had taken its toll. It might be a bug that causes slight discolouration or a lack of rain at the time the corn ears pop; it's easy to explain the changes to our food if one takes the time to listen. The farmer agreed to sell the corn to Robert, but he sent it in unmarked boxes and refused to put his name to it. But if you put your name to the flaw, Robert reckons, 'that's when you get to the source of your true identity'.

Why is it so hard to acknowledge failure when there's so much potential in that act of telling?

To farm is to be at the mercy of nature; that is true the world over. To own the identity of 'farmer' is to embrace the failure often caused by markets in flux, the unbending will of nature, and the seasons, which are now so unreadable, so prone to change. To be a farmer in the age of global warming is to be very far from perfection. But in the embrace of one's flaws, the potential for seismic change and beauty emerges.

I kept asking for failure stories and farmers offered them up, one by one. They all have them.

Wayne Langford farms on the northern tip of New Zealand's South Island, around the azure bays with unstepped golden sand where the Langford name is known. They are sixth-generation farmers and landowners here. And for a while the family was focused on buying up other dairy farms to grow the operation. Wayne and his wife, Tyler, were the family entrepreneurs — always sent off to run the latest purchase. And then he quit. The family and the farms. It was a 'bit of a dust-up', he tells me. Now he and his wife milk 250 cows on a small property, and his focus is regularly elsewhere. His demeanour is jovial, but it wasn't always.

Robert Pekin once told me dairy farmers are the angriest of people. He was talking from the experience of losing his family property when Australia deregulated the dairy industry in 1999, removing government-mandated pricing for drinking milk. He was living on a property he'd paid too much for, putting all his energy into producing as much milk as possible: more inputs, more fertiliser, more cows, less money. He didn't consider himself a food producer; he just wanted to be the best farmer around. But it was anger, not accolades, that descended.

Robert left farming. And a decade later he founded Food Connect. Wayne left the family farming enterprise and later, with fellow dairy farmer Siobhan O'Malley, founded Meat the Need, a charity that sources cattle from New Zealand farmers and milk from the Māori-owned dairy company Miraka, to keep over-subscribed food banks across the country in steady supply of premium red meat and dairy.

To use food to care for others does something profound to a person's sense of worth.

In an industry rooted in conservative ideology, Wayne is by his own admission a 'bit socialistic' and prone to ideas that are 'kind of unicorns and rainbows'. Meat the Need is a unicorn idea born from a feeling of obligation to his community. It is a very simple proposition, but as Wayne reminded me, it required the right person at the right time to drive it. It had to be farmer-led, he told me, for others to buy into the vision. Via Meat the Need, Wayne has chosen to lift those most in need of the food he and his cohort produce. It's enough for now, but not for good. Wayne says there is still so much more work to do to provide affordable food for those New Zealanders who each week are struggling to put food on the table.

Wayne holds the position of Dairy Industry Group Chairperson for New Zealand's Federated Farmers, an advocacy group that started as a union back in 1899. But it's as a concerned farmer and Fonterra shareholder that he seeks out the ear of influential dairy industry power brokers in order to float a rainbow beam their way. He wonders, for example, why the Fonterra cooperative doesn't better utilise its factories to produce dairy products at a price the domestic market can easily afford. There is an imbalance, currently. And an expectation that New Zealanders should be willing to pay the same for local cheese or prime steak as those who buy it in London or San Francisco.

A kilogram of edam, colby, tasty, or mild cheddar cheese costs around NZ$15. A 500 g block of butter is $6. A litre of milk $2.50. The dairy industry produces around 21 billion litres of milk each year and contributes NZ$7.8 billion to the New Zealand GDP; it is a sector of great wealth that currently sits apart from the many New Zealanders going hungry. Access to affordable food is a human right, but it's not a legislated corporate responsibility. So, it's more likely to fall to individuals to create a system that privileges fairness and affordability — to remake the food system for community.

In Australia, Robert Pekin is one such disruptor. Wayne is another. And the catalyst for change might just be great personal loss. Some will see it as failure; others will be inspired by the resilience to keep going. For Robert and Wayne, losing a connection with long-held family farms was cause for introspection. For me, it was a miscarriage that sparked the questioning.

There is a reckoning to be had — not with each other, but with ourselves.

We have been told so many times in the past few years that we all need to act to curb climate change. So many of us have tried to change, and we started at the supermarket because we were told it was the right thing to do. Change your diet, they said. That is the solution. But how many of you who started failed as I have failed, over and over again? As Pekin likes to say, the world is not perfect. But if there is one thing I've learned above all else, it's that failure is not a flaw. Fail and fail again, but don't give up trying to change. If you do, we won't survive.

9.

Don't give up

There is a fisherman who operates from the southernmost port of New Zealand who has a son named Forest, who like his dad has a passion for the sea. If Forest takes to the ocean to catch fish, he'll be a fourth-generation fisher — the fourth generation that knows how to listen to the ocean.

Once a commercial operator for a large New Zealand seafood company, this fisher is now the captain of his own boat, called *Gravity*. Painted below the boat name on the bridge is a skull and crossbones: a pirate symbol to herald the presence of someone who doesn't always play by the rules. I wanted to talk to Nate Smith, this fisher, because he's an oddity, a sole operator who is trying to shift an industry to do more than simply brand itself sustainable. And he's doing it from the bracing, swirling waters of the Foveaux Strait, the expanse that separates the South Island of New Zealand from the dot of land most people forget: Stewart Island. Beyond is Antarctica.

Nate's business is simple, but it's profound in its difference from conventional fishing and food production. He catches on order, only what's in season, and then he ships boxes of fish killed instantly by a process called Ikejime (a swift pierce of a rod into the brain as perfected by Japanese fishers) to restaurants, lodges, and kitchens around the country. No middleman, no fishing over quota, all traceable.

Transparency in the supply chain is paramount, and his customers know exactly where the fish is from and how it got to their kitchen, because they all have his phone number. This is fishing without destroying, he says. With a thoughtfulness cultivated by hours alone on the sea, he seems first confused, then inflamed at the injustice of our collective situation: 'We are nature ourselves,' he says. 'We have a connection to the land that goes all the way through us and if we don't look after these things … it's simple. We won't survive.'

We won't survive.

His son, my daughter, and their children will struggle through increasing heatwaves, droughts, fierce storms, changes in growing seasons, and floods. Biodiversity will plummet even further as species suited to the climate of yesteryear disappear. If we don't act now, we'll hit 3 degrees warming, according to the world's leading climate scientists. To start is to figure out how to listen to the ocean and look through the land, gestures of quiet activism that are potentially more potent than any protests on the streets.

It's been years since I left Sydney, resigned from my role as Managing Editor for BuzzFeed, packed a shipping crate with my furniture and possessions, and convinced my Australian husband to move countries with me so we could become farmers. In New Zealand. Land of milk and honey. I was going to be a food producer. Live on the land. Be good to the environment. Be a farmer.

Then the animal rights activist group PETA told me there's no such thing as a meat-eating environmentalist.

But they were wrong.

—

The afternoons have started to heat up; it's November and the farm is flush with grasses. Some we've purposefully introduced; others have appeared out of the seed bank, pushing shoots up amid the rye grass to announce their presence. Wild radish, common plantain, red clover, lesser trefoil, and comfrey — a medicinal plant for us that we can pick to make a salve for bruised limbs, a tonic to fertilise vegetables, and a nutritional treat for our animals. In the early morning, as the fog lifts off the lake, we pack a day bag into the farm truck — diesel not electric, yet — and drive out of the farm gates heading toward Hawke's Bay, and my mum's family farm.

In the back seat my daughter bellows the sound of a cattle beast, a generous 'MOOOOO', as she spots animals in the paddocks down our driveway, and then out on the road. She keeps it up until she falls asleep 40 minutes later, as we head out into the tussock country of the Rangitaiki plains.

Not long ago, an environmental advocacy group presented a research paper to the Hawke's Bay regional council, suggesting that the area was at risk of desertification if little was done to mitigate warming.[1] They encouraged the region to invest in water reservoirs, to at least ensure steady supply of water to the cities and towns that dot the coast of Hawke's Bay. It's a region in flux. Two weeks before we travel, one of the larger towns, Napier, has flooded in a downpour that pushed 242.4 millimetres of rain through the tired stormwater systems within 24 hours. It was the biggest rain event since 1963. It's like Sydney weather, Pat and I say. Drought followed by flood. But New Zealand isn't prepared for the epic weather events that Australia has long endured. Not yet, anyway.

We enter Hawke's Bay via a natural gateway at the top of the Ahimanawa Range, and the land and coastline splays in front of us. Vineyards, orchards, and cropping, sheep and cattle farms line the road as we head south, past the flooded town and towards the flat plains of southern Hawke's Bay, where my cousins farm the old family property.

I've come because I've learned that Pat and I are not entirely alone.

When we arrive, two generations of cousins settle in along the bench seat, next to the window that looks out over cascading manicured lawns. They talk of regenerative farming, but in their words is the weight of devotion to the task: they hold the fate of the future generations on their collective palm.

My elder cousin says, with some sadness, that he feels the chasm is so wide now between people who think they know what happens on farms and those who do the farming. His big, coarse hands are clasped in front of him, elbows resting on the table, head bent for just a moment in contemplation. The younger cousins have been reading the reports about climate change, and summer after summer of dry hot weather is enough warning for them to act. Like us, they're changing. The elder cousin mostly sits quietly listening to his sons, nodding as they outline the changes.

Our conversation takes on a similar pattern to many I've had over the past two years with farmers who believe they are part of the solution. Like those who've already been feeding their dairy cows a methane suppressant seaweed supplement for years; turning vast rows of compost instead of spreading synthetic fertiliser; or creating a vast Australian farmscape dotted by solar panel clusters, under which the arid landscape grows grasses for livestock, or vegetables and crops kept moist from the condensation that runs off the panels, so that food continues to grow under the blast of the 42-degree summer days.

The great food transformation that's been heralded by food policy experts has called for everyone along the chain to act, but many of the actions are small and subtle, and they happen on family-run farms that aren't issuing press releases or annual reports. So, the changes go unnoticed. But they're there. There are people farming who know what's at stake and they want to be the solution to warming, not the problem. They want to feed people, not a commodity system.

We leave after lunch and drive back home, where we find a clutch of 20 quails that live in the stand of trees at the back of the house calling to each other. A melodic hoot from one will bring the others running in towards the vegetable garden.

I hear the calls of protest loudly in the most unexpected places now. It's coming from our tree stand, but also boardrooms and courtrooms, where judges are ruling in favour of plaintiffs trying to change the world order. In Australia, eight teenagers, led by Anj Sharma and an octogenarian nun named Sister Brigid Arthur, brought a case against the government, arguing for an injunction against a newly proposed Whitehaven Coal mine in New South Wales by claiming that then environment minister Sussan Ley had a common-law duty of care to not act in a way that might cause future harm to young Australians. Justice Mordecai Bromberg agreed with Sharma — the duty of care exists — but the then minister initiated an appeal, and the fight continues.[2]

In the Netherlands, a Dutch Court forced Royal Dutch Shell to cut its company emissions by 45 per cent before 2030, stating the company was violating the Netherlands' civil code through activities that endanger human lives.[3] In the United States, fossil fuel giants ExxonMobil and Chevron faced shareholder rebellions from climate activists and dissatisfied institutional investors over inaction on climate mitigation.[4] An activist hedge fund staged a coup that resulted in it capturing two board seats at Exxon, from where it can help drive the oil company towards a greener strategy. Protest is cannibalising the host from the inside out.

The work Pat and I do — to cut fossil fuels from our farming system, to bring diversity back within the boundary, to nurture worms and fungi and monarch butterflies, to not plough the land, to not grow winter crops, to kill animals before their collective footprint is too much for the land to bear, to eat less beef — is as stinging a protest as the hedge fund capturing those board seats.

People are watching; those most fearful are waiting for us to fail. And we will fail, but we won't give up. Only then, when we try again and again, will others move into a space of opportunity that famers like us are creating. We don't need to change everyone around us; we only need to create a space for those ahead of us. Some will be left behind. Some, like my dad, won't come.

He visits occasionally, always bringing with him an offering: newly pickled gherkins from his kitchen, plums from his tree, exaggerated chard leaves made big with heavy application of sheep droppings on his small vegetable plot. We sit outside overlooking the lake, where once he'd shoot ducks. And he'll ask a few questions, about the cattle, mostly. Then he'll leave. The act of leaving is a gracious gesture of allyship without having to ever utter the words 'I'm with you'. He leaves so that Pat and I can do the work that he won't recognise.

For Christmas, my sister gives him a copy of the Edmonds Cookery Book; a collection of 'traditional New Zealand recipes'. Through food he is pitching backward to a time when we'd dock lambs as a family, pick clingstone peaches from the tree under the warm summer sun, eat roasts on Sunday and leftover meat fritters on Monday. Dousing his meals with a heavy layer of nostalgia and comfort. This cooking is a new habit of his that reminds me of an essay French artist Daniel Buren wrote, in which he reflected on the nature of the artist's studio and the role it plays to enclose and cocoon a work of art. He wrote that an artwork is closest to its own reality only when in the studio. And from that moment, it never ceases to distance itself from that reality. 'It may become what even its creator had not anticipated,' writes Buren. 'Serving instead, as is usually the case, the greater profit of financial interests and the dominant ideology.'

The same can be said for food. On the land from which it comes, it has a defined reality as object or animal — a corn cob, a carrot, a cow — but in the hands of humans, it is ascribed myriad meanings: nostalgic totem, lunch, nutrients, sustenance, history, ritual, milk, cheese, whey, baby milk powder, fructose syrup, starch, soft drink, chips flavouring, exports, income, jobs, revenue, ideology, and activism. Use it as a tool with which to save the planet, though, and it's a thing ascribed with too much meaning.

Food has the power to transport and nurture, if you give it the room to do so. If it's reduced to an equation, a sum of footprints, methane and carbon dioxide equivalents, then we will fail at our task, because that process removes all trace of culture, care, and joy — the very things we need most to get us through. If it's used to shun, to label, to accuse someone else of inaction, we will fail. For that is a puny act that ruptures community, the very thing we need to rebuild.

If you've come this far and are waiting for advice on how to act, what small gesture you can make to do your bit for the environment, you may find my response overwhelming. Because, in my experience, it's change everything, fail often, make sacrifice after sacrifice. Reduce your red meat consumption, don't waste food, support renewable energy initiatives, vote for politicians who aren't backed by fossil fuel money, stand for local council and block proposals for intensified farm conversions, plant native trees, pool funds and stage a shareholder coup against a major polluter, buy fruit and vegetables from food hubs, support small, farmer-focused food retailers, travel less, walk more … change everything.

I tried everything, but I was just spiralling. And then I stopped. When Pat and I arrived in New Zealand to farm, we slammed against each other to build a business, a life, something bigger than ourselves. It fractured something in us, but not irreparably. Someone once told me that between two people a space exists that needs to be nurtured and fed for the relationship to flourish. I don't need to convince Pat to be more like me, nor I more like him. The feeding will ebb and wane, but someone will always nurture that space, and the result is togetherness.

It's an analogy that can be applied to our collective will to curb climate change. We don't need to mirror each other in ambition, ethics, and action, but we all need to keep feeding the space that will slow the warming. The 'how' is up to you, but the doing of it will make the collective stronger. Together but apart. Profound activism can be practised when you reattribute meaning to the time spent doing things that feed this space. Things previously thought of as insignificant, too small to matter — like the cooking of a meal from scratch for your children.[5] Or the decision to be less busy, to do less. I don't need to convert or convince you to be more like me. Nor I like you.

The life I lived in Sydney, and London and Melbourne before that, kept me away from death. Until my miscarriage, when the difference between a beating heart and silence was made clear. I succumbed to it for years, seeing death in everything, thinking the planet was humming its last note. But even this is a moment of transition — part of a cycle that provides for new life, or a new type of life.

A harrier hawk raids a stoat nest and drops the nude kittens from a great height, leaving them exposed and lifeless on the hill above the lake. Such is life.

Grasses and vegetables flourish and drop, decomposing into the dank soil. Soon to give off the perfume of chocolate cake.

The beat and hum of wings, the rumble of distant clouds, the pulse from a subterranean web of roots, taps, nodules, and organisms that allows me to coax food from the land is the death-into-life cycle now loud around me. It sounds like music.

'Photosynthesis is happening!' Pat bellows. She giggles. 'Ssssh Daddy. Sssssh.'

'Are you done yet?' he asks, peeking one eye through the crack in the office door.

No, I say. I'm not done.

Acknowledgements

This book started as an excuse for my dearest friend, Naima Brown, and me to talk daily after I left Sydney and she stayed. After the conversations, we wrote television-show pitches about food and farming, documentary treatments, and then, finally, a development funding application for a podcast that would become *A Carnivore's Crisis* — an eight-part series I made with Naima and the wonderful cook, broadcaster, and writer Rachel Khoo for Audible Australia. Weeks after we released the show, Naima asked me to work with her again on another commission. I said, 'No, thanks. I'm going to stay put; I think this it for me.' I realised then I'd found my beat writing about food and farming. It is the most satisfying work I've done in my career to date. I am grateful to Naima Brown for having those early conversations, for collaborating, showing up to record interviews, for forgiving me when I made mistakes, reading early drafts, and believing in my vision.

I wrote this book on farmland that flanks a waterway that's significant for *Te Kapa o Te Rangiita* and I feel a responsibility to nurture and care for this taonga in all I do, whether that be in my writing or farming practices. I am grateful to specific members of this hapu for their advice and guidance on how to best do this work.

Some of the ideas in this book have been tested on other platforms and refined by colleagues including Rachel Morris and Rebecca Stevenson; thank you both for your insights. And thank you also to the friends and colleagues who agreed to read chapters and early drafts: Anja MacDonald, Anna Sulan Masing, Pip Cummings, and most especially Brad Esposito, who offered the most insightful summary of my relationship with Pat: 'He works, you think. He gets angry, you ask why are we angry?' Exactly. Thankfully, Pat finds all my thinking and questioning charming (most of the time).

I am indebted to the great many people who agreed to speak to me for this book on and off the record. There is a complex system in which farming knowledge is shared and once you're accepted into the network the ideas, information, stories, and myths start following. Thanks most especially to the White family, the Andersons, Hamish, John and the farm bros, and supportive ag women of Twitter for nudging me in the right direction and challenging me along the way.

I am indebted to my mum, Deb, for handing over the tranche of research, letters, journals, and photos that guided my attempt to tell the White family's arrival story. And to my dad Roger and uncle Graeme for providing so much rich material.

But there are two people especially who helped me bring this story to life and I am grateful to them both. My editor at Scribe, Marika Webb-Pullman, saw the story buried deep under the research and I am in awe of her skill. My agent, Grace Heifetz from Left Bank Literary, said yes to my 'Big Idea' and I can't thank her enough.

And finally thank you to my husband, Pat, and daughter, Clara: you are both the beating heart of this story and the reason I keep going.

Bibliography

Books

Burkeman, O. (2021). *Four Thousand Weeks: time and how to use it*. Vintage Publishing.

Carolan, M. (2018). *The Real Cost of Cheap Food* (2nd ed.). Routledge.

Carson, R. (1962). *Silent Spring*. Houghton Mifflin.

Evans, M. (2019). *On Eating Meat*. Murdoch Books.

Evans, M. (2021). *Soil*. Murdoch Books.

Fiso, M. (2020). *Hiakai: modern Māori cuisine*. RHNZ Godwit.

Gammage, B. (2011). *The biggest estate on earth: how Aborigines made Australia*. Allen & Unwin.

Goodman, D., & Redclift, M. (1991). *Refashioning Nature: food, ecology and culture*. Routledge.

Grace, A. (2018). *This Naked Mind*. Harper Collins.

Gray, L. (2018). *The Ethical Carnivore: my year killing to eat*. Bloomsbury.

Hesser, L. (2019). *The Man Who Fed the World*. Righter's Mill Press LLC.

Holt-Giménez, E., & Patel, R. (2010). *Food Rebellions!: forging food sovereignty to solve the global food crisis*. Fahamu.

Lind, C.A. (2013). *Till the Cows Came Home: inside the battles that built Fonterra*. Steele Roberts.

Little, A. (2019). *The Fate of Food*. One World.

Macgregor, M. (1981). *Petticoat Pioneers: North Island women in the colonial era*. A.H. & A.W. Reed.

Nussbaum, M.C. (2011). *Creating Capabilities: the human development approach*. First Harvard University Press.

Parham, B.E.V. & Healy, A.J. (1976). *Common Weeds in New Zealand*. A.R. Shearer.

Parker, C., Carey, R., & Scrinis, G. (2019). The Consumer Labelling Turn in Farmed Animal Welfare Politics: from the margins of animal advocacy to mainstream supermarket shelves. In M. Phillipov & K. Kirkwood (Eds.), *Alternative Food Politics: from the margins to the mainstream* (pp. 193–215). Routledge (Critical Food Series).

Pascoe, B. (2014). *Dark Emu*. Magabala Books.

Phillipov, M., & Kirkwood, K. (Eds.). (2019). *Alternative Food Politics: from the margins to the mainstream*. Routledge.

Sanghera, S. (2021). *Empireland: how modern Britain is shaped by its imperial past*. Viking.

Savory, A. (2016). *Holistic Management* (3rd ed.). Island Press.

Schulman, S. (2018). *Conflict Is Not Abuse: overstating harm, community responsibility and the duty of repair*. Arsenal Pulp Press.

Stuart, T. (2009). *Waste: uncovering the global food scandal*. Penguin.

Tree, I. (2019). *Wilding: the return of nature to a British farm*. Pan Macmillan.

Tudge, C. (2004). *So Shall We Reap*. Penguin.

Vialles, N. (1994). *Animal to Edible*. (J.A. Underwood, Trans.). Cambridge University Press. (Original work published 1987).

Wallace, N. (2014). *When the Farm Gates Opened: the impact of Rogernomics on rural New Zealand*. Otago University Press.

Wilkerson, I. (2020). *Caste: the origins of our discontents*. Random House.

Papers, reports, and academic studies

Allen, M. (2015). *Short-Lived Promise? the science and policy of cumulative and short-lived climate pollutants*. Oxford Martin Policy Paper, University of Oxford. https://www.oxfordmartin.ox.ac.uk/downloads/briefings/Short_Lived_Promise.pdf

Allen, M., Cain, M., Lynch, J., & Frame, D. (2018). *Climate Metrics for Ruminant Livestock, Oxford Martin Programme on Climate Pollutants*. Oxford Martin School. https://www.oxfordmartin.ox.ac.uk/downloads/reports/Climate-metrics-for-ruminant-livestock.pdf

Altieri, M., & Pengue, W. (2006). GM Soybean: Latin America's new colonizer. *Grain*. https://www.grain.org/article/entries/588-gm-soybean-latin-america-s-new-colonizer

Anthony, T. (2007). Criminal Justice and Transgression on Northern Australian Cattle Stations. In I. Macfarlane & M. Hannah (Eds.), *Transgressions: critical Australian Indigenous histories* (Vol. 16, pp. 35–62). ANU Press. https://doi.org/10.22459/T.12.2007.03

Auckland City Mission. (2019). *Shining the Light on Food Insecurity in Aotearoa: Auckland City Mission's call to action*. https://www.aucklandcitymission.org.nz/wp-content/uploads/2019/10/org07307-ACM-world-food-day-A5_web.pdf

Austin, E.J, Deary, I.J, Edwards-Jones, G., & Arey, D. (2005). Attitudes to Farm Animal Welfare: factor structure and personality correlates in farmers and agriculture students. *Journal of Individual Differences*, 26(3), 107–120.

Bartram, D.J., & Baldwin, D.S. (2010). Veterinary Surgeons and Suicide: a structured review of possible influences on increased risk. *Veterinary Record*, 166, 388–397. https://doi.org/10.1136/vr.b4794

Beck, K.L., Conlon, C.A., Kruger, R., Heath, A-L.M., Matthys, C., Coad, J., & Stonehouse, W. (2012). Iron Status and Self-Perceived Health, Well-being, and Fatigue in Female University Students Living in New Zealand. *Journal of the American College of Nutrition*, 31(1), 45–53.

Bell, C. (1997). The 'real' New Zealand: rural mythologies perpetuated and commodified. *The Social Science Journal*, 34(2), 145–158. https://doi.org/10.1016/S0362-3319(97)90047-1

Bell, C. (2011). Farmers' Markets: commoditising New Zealand rural identity myths. *Przestrzeń Społeczna* (Social Space), 1/2, 57–74.

Bowden, M. (2020). Understanding Food Insecurity in Australia (CFCA Paper No. 55). https://aifs.gov.au/cfca/publications/understanding-food-insecurity-australia

Bylsma, L.C., & Alexander, D.D. (2015). A Review and Meta-Analysis of Prospective Studies of Red and Processed Meat, Meat Cooking Methods, Heme Iron, Heterocyclic Amines and Prostate Cancer. *Nutrition Journal*, 14, 125.

Campbell, H. (2006). Consultation, Commerce and Contemporary Agri-Food Systems: ethical engagement of new systems of governance under reflexive modernity. *The Integrated Assessment Journal*, 6(2), 117–136.

Campbell, H. (2009). Breaking New Ground in Food Regime Theory: corporate environmentalism, ecological feedbacks and the 'food from somewhere' regime? *Agriculture and Human Values*, 26(4), 309–319.

Carey, R., Parker, C., & Scrinis, G. (2017). Capturing the Meaning of 'Free Range': the contest between producers, supermarkets and consumers for the higher welfare egg label in Australia. *Journal of Rural Studies*, 54, 266–275. https://doi.org/10.1016/j.jrurstud.2017.06.014

Carlson, K.M., Gerber, J.S, Mueller, N.D, Herrero, M., MacDonald, G.K., Brauman, K.A, Havlik, P., O'Connell, C.S., Johnson, J.A., Saatchi, S., & West, P.C. (2016). Greenhouse Gas Emissions Intensity of Global Croplands. *Nature Climate Change*, 7, 63–68.

Carter, K., Lanumata, T., Kruse, K., & Gorton, D. (2010). What Are the Determinants of Food Insecurity in New Zealand and Does This Differ for Males and Females? *Australian and New Zealand Journal of Public Health*, 34, 602–608. https://www.otago.ac.nz/wellington/otago020409.pdf

Cook, J.D, Skikne, B.S., & Baynes, R.D. (1994). Iron Deficiency: the global perspective. *Advances in Experimental Medicine and Biology*, 356, 219–228.

Cordain, L., Eaton, S.B., Brand Miller, J., Mann, N., & Hill, K. (2002). The Paradoxical Nature of Hunter-Gatherer Diets: meat-based, yet not atherogenic. *European Journal of Clinical Nutrition*, 56(1), S42–S52.

Craig, J.L. (1974). *The Social Organization of the Pukeko, Porphyrio porphyrio melanotus, Temminck, 1820* [Unpublished doctoral thesis]. Massey University. https://mro.massey.ac.nz/handle/10179/3613

Dillard, J. (2008). Slaughterhouse Nightmare: psychological harm suffered by slaughterhouse employees and the possibility of redress through legal reform. *Georgetown Journal on Poverty Law and Policy*, 15(2), 391–408.

Fan, S., Cho, E.E., & Rue, C. (2017). Food Security and Nutrition in an Urbanizing World: a synthesis of the 2017 Global Food Policy Report. *China Agricultural Economic Review*, 9(2), 162–168.

Foley, J.A, Ramankutty, N., Brauman, K.A., Cassidy, E.S., Gerber, J.S., Johnston, M., Mueller, N.D., O'Connell, C., Ray, D., West, P., Balzer, C., Bennett, E., Carpenter, S., Hill, J., Monfreda, C., Polasky, S., Rockström, J., Sheehan, J., Siebert, S., & Zaks, D.P.M. (2011). Solutions for a cultivated planet. *Nature*, 478, 337–342.

Fontefrancesco, M.F. (2019). Food Commodity Market: history and impact of food trading toward SDG2. In W. Leal Filho, A.M. Azul, L. Brandli, P.G. Özuyar, & T. Wall (Eds.), *Zero Hunger: Encyclopedia of the UN Sustainable Development Goals* (pp. 304–312). Springer, Cham. https://doi.org/10.1007/978-3-319-95675-6_13

Friedmann, H. (2005). From Colonialism to Green Capitalism: social movements and the emergence of food regimes. In F.H. Buttel & P.D. McMichael (Eds.), *New Directions in the Sociology of International Development*, 11 (Research in Rural Sociology and Development) (pp. 227–264). Emerald Publishing Limited. doi:10.1016/S1057-1922(05)11009-9

Garnett, T., Godde, C., Muller, A., Röös, E., Smith, P., de Boer, I.J.M., zu Ermgassen, E., Herrero, M., van Middelaar, C., Schader, C., & van Zanten, H. (2017). *Grazed and Confused? ruminating on cattle, grazing systems, methane, nitrous oxide, the soil carbon sequestration question – and what it all means for greenhouse gas emissions*. Food Climate Research Network, University of Oxford. https://www.oxfordmartin.ox.ac.uk/downloads/reports/fcrn_gnc_report.pdf

Gómez, M.I., Barrett, C.B., Pinstrup-Andersen, P., Raney, T., Meerman J., Croppenstedt, A., Lowder, S., Carisma, B., & Thompson, B. (2013). *Post-Green Revolution Food Systems and the Triple Burden of Malnutrition* (ESA Working Paper No. 13–02). Agricultural Development Economics Division Food and Agriculture Organization (FAO) of the United Nations (UN). www.fao.org/economic/esa

Gow, N.G. (2014). New Zealand Government's Involvement in Agriculture – the Road to Non-Sustainability. IFMA, 16 (Theme 1). https://researcharchive.lincoln.ac.nz/bitstream/handle/10182/418/07_N_Gow.pdf

Graham, R. (2017). *The Lived Experiences of Food Insecurity Within the Context of Poverty in Hamilton, New Zealand* [Unpublished doctoral thesis]. Massey University. https://mro.massey.ac.nz/handle/10179/13001

Grandin, T. (1998). Fast Food Chains Audit Animal Handling Practices. *Meat & Poultry*, 57.

Grelet, G-A., Lang, S., Merfield, C., Calhoun, N., Robson-Williams, M., Horrocks, A., Dewes, A., Clifford, A., Stevenson, B., Saunders, C., Lister, C., Perley, C., Maslen, D., Selbie, D., Tait, P., Roudier, P., Mellor, R., Teague, W.R., Gregory, R., ... Langford, W. (2021). *Regenerative Agriculture in Aotearoa New Zealand– research pathways to build science-based evidence and national narratives* [White paper]. Manaaki Whenua

Landcare Research. https://ourlandandwater.nz/wp-content/uploads/2021/02/ Grelet_Lang_Feb-2021_Regen_Ag_NZ_White_ePaper.pdf

Kennedy, A., Adams, J., Dwyer, J., Muhammad, A.R., & Brumby, S. (2020). Suicide in Rural Australia: are farming-related suicides different? *International Journal of Environmental Research and Public Health*, 17(6), 2010. https://doi.org/10.3390/ ijerph17062010

Khoury, C.K., Bjorkman, A.D., Dempewolf, H., Ramirez-Villegas, J., Guarino, L., Jarvis, A., Rieseberg, L.H., & Struik, P.C. (2014). Increasing Homogeneity in Global Food Supplies and the Implications for Food Security. *PNAS*, 109(31), 12302–12308. https://www.pnas.org/content/111/11/4001

Kidd, B., Mackay, S., Vandevijvere, S., & Swinburn, B. (2021). Cost and Greenhouse Gas Emissions of Current, Healthy, Flexitarian and Vegan Diets in Aotearoa (New Zealand). *BMJ Nutrition, Prevention & Health*. doi:10.1136/bmjnph-2021-000262

Kinley, R.D., de Nys, R., Vucko, M.J., Machado, L., & Tomkins, N.W. (2016). The Red Macroalgae *Asparagopsis taxiformis* Is a Potent Natural Antimethanogenic That Reduces Methane Production During *in vitro* Fermentation with Rumen Fluid. *Animal Production Science*, 56, 282–289. https://doi.orgz10.1071/AN15576

Kinnunen P., Guillaume, J.H.A., Taka, M., D'Odorico, P., Siebert, S., Puma, M.J., Jalava, M., & Kummu, M. (2020). Local Food Crop Production Can Fulfil Demand for Less Than One-Third of the Population. *Nature Food*, 1, 229–237. https://doi.org/10.1038/ s43016-020-0060-7

Hafting, J., Critchley, A., Cornish, M., Hubley, S., & Archibald, A. (2012). On-Land Cultivation of Functional Seaweed Products for Human Usage. *Journal of Applied Phycology*, 24, 385–392. doi:10.1007/s10811-011-9720-1

Harker-Schuch, I., Lade, S., Mills, F., & Colvin, R. (2021). Opinions of 12 to 13-Year-Olds in Austria and Australia on the Concern, Cause and Imminence of Climate Change. *Ambio*, 50, 644–660. https://doi.org/10.1007/s13280-020-01356-2

Harwatt, H. (2019). Including Animal to Plant Protein Shifts in Climate Change Mitigation Policy: a proposed three-step strategy. *Climate Policy*, 19(5), 533–541. doi :10.1080/14693062.2018.1528965

Hegab, A., Fayed, M.T.B., Hamada, M., & Abdrabbo, M. (2014). Productivity and Irrigation Requirements of Faba-Bean in North Delta of Egypt in Relation to Planting Dates. *Annals of Irrigation Sciences*, 59, 185–193. doi:10.1016/j.aoas.2014.11.004

Lambert, M.G., Roberts, A.H.C, & Morton, J.D. (2012). Nitrogen Use on Hill *Country: lessons from the national Wise Use of Nitrogen focus farm project*. Farmed Landscapes Research Centre, Lincoln University. http://flrc.massey.ac.nz/workshops/12/ Manuscripts/Lambert_2012.pdf

Lang, T., Millstone, E., & Marsden, T. (2017). *A Food Brexit: time to get real*. University of Sussex Science Policy Research Unit. https://www.sussex.ac.uk/webteam/gateway/ file.php?name=foodbrexitreport-langmillstonemarsden-july2017pdf.pdf&site=25

Ledgard, S.F, Penno, J.W., & Sprosen, M.S. (1997). Nitrogen Balances and Losses on Intensive Dairy Farms. *Proceedings of the New Zealand Grassland Association*, 59, 49– 53. https://www.grassland.org.nz/publications/nzgrassland_publication_532.pdf

Farm

Leibler, J.H., Janulewicz, P.A., & Perry, M.J. (2017). Prevalence of Serious Psychological Distress Among Slaughterhouse Workers at a United States Beef Packing Plant. *Work*, 57(1), 105–109. doi:10.3233/WOR-172543

Lynch, J., & Pierrehumbert, R. (2019). Climate Impacts of Cultured Meat and Beef Cattle. *Frontiers in Sustainable Food Systems*, 3, 5. https://doi.org/10.3389/fsufs.2019.00005

Ma, Q., Kim, E.Y., Lindsay, E.A., & Han, O. (2011). Bioactive Dietary Polyphenols Inhibit Heme Iron Absorption in a Dose-Dependent Manner in Human Intestinal Caco-2 Cells. *Journal of Food Science*, 76(5), H143–H150. https://doi.org/10.1111/j.1750-3841.2011.02184.x

Maseyk, F.J.F., Small, B., Henwood, R.J.T, Pannell, J., Buckley, H.L., & Norton, D.A. (2021). Managing and Protecting Native Biodiversity On-Farm – What Do Sheep and Beef Farmers Think? *New Zealand Journal of Ecology*, 45(1), 3420. https://dx.doi.org/10.20417/nzjecol.45.1

Mazzetto, A., Falconer, S., & Ledgard, S. (2021). *Mapping the Carbon Footprint of Milk for Dairy Cows*. Dairy NZ and Ag Research. https://www.dairynz.co.nz/media/5794083/mapping-the-carbon-footprint-of-milk-for-dairy-cows-report-updated.pdf

Montgomery, J. (2013). *Animal Welfare and Animal Rights: a war of words with casualties mounting*. Animal Welfare Council. https://www.animalwelfarecouncil.org/?page_id=473

Moriarty, E.M., Karki, N., Mackenzie, M., Sinton, L.W., Wood, D.R., & Gilpin, B.G. (2011). Faecal Indicators and Pathogens in Selected New Zealand Waterfowl. *New Zealand Journal of Marine and Freshwater Research*, 45(4), 679–688. doi:10.1080/00288330.2011.578653

National Trust. (2020). *Interim Report on the Connections between Colonialism and Properties Now in the Care of the National Trust, Including Links with Historic Slavery*. https://nt.global.ssl.fastly.net/documents/colionialism-and-historic-slavery-report.pdf

Nawroth, C., Langbein, J., Coulon, M., Gabor, V., Oesterwind, S., Benz-Schwarzburg, J., & von Borell, E. (2019). Farm Animal Cognition—Linking Behavior, Welfare and Ethics. *Frontiers in Veterinary Science, 6*. https://www.frontiersin.org/article/10.3389/fvets.2019.00024

Newman, N., Fletcher, R., Schulz, A., Nielsen, S., & Kleis, R. (2020). *Reuters Institute Digital News Report 2020*. Reuters Institute for the Study of Journalism. https://reutersinstitute.politics.ox.ac.uk/sites/default/files/2020-06/DNR_2020_FINAL.pdf

Pereira, L., Calderón-Contreras, R., Norström, A., Espinosa, D., Willis, J., Guerrero, L.L, & Pérez, A.O. (2019). Chefs as Change-Makers from the Kitchen: indigenous knowledge and traditional food as sustainability innovations. *Global Sustainability*, 2, E16. doi:10.1017/S2059479819000139

Pereira, L., Drimie, S., Maciejewski, K., Tonissen, P.B., & Biggs, R.O. (2020). Food System Transformation: integrating a political–economy and social–ecological approach to regime shifts. *International Journal of Environmental Research and Public Health, 17*, 1313. https://doi.org/10.3390/ijerph17041313

Persson, K., Felicitas, S., Neitzke, G., & Kunzmann, P. (2020). Philosophy of a 'Good Death' in Small Animals and Consequences for Euthanasia in Animal Law and Veterinary Practice. *Animals*, 10(1), 124. https://doi.org/10.3390/ani10010124

Pingali, P.L. (2012). Green Revolution: toward 2.0. *PNAS*, 109(31), 12302–12308.

Porcher, J. (2006). Well-Being and Suffering in Livestock Farming: living conditions at work for people and animals. *Sociologie du Travail*, 48, 56–70.

Potter, H., Lundmark, L., & Röös, E. (2020). *Environmental Impact of Plant-Based Foods – data collection for the development of a consumer guide for plant-based foods* (Report 112). Swedish University of Agricultural Sciences, NL Faculty/Department of Energy and Technology. https://pub.epsilon.slu.se/17699/1/Report112.pdf

Richards, E., Signal, T., & Taylor, N. (2013). A Different Cut? comparing attitudes toward animals and propensity for aggression within two primary industry cohorts—farmers and meatworkers. *Society & Animals*, 21, 395–413. https://www.animalsandsociety.org/wp-content/uploads/2016/05/richards.pdf

Rush, E., & Obolonkin, V. (2020). Food Exports and Imports of New Zealand in Relation to the Food-Based Dietary Guidelines. *European Journal of Clinical Nutrition, 74*, 307–313. https://doi.org/10.1038/s41430-019-0557-z

Springmann, M., Clark, M., Mason-D'Croz, D., Wiebe, K., Bodirsky, B.L., Lassaletta, L., de Vries, W., Vermeulen, S.J., Herrero, M., Carlson, K.M., Jonell, M., Troell, M., DeClerck, F., Gordon, L.J., Zurayk, R., Scarborough, P., Rayner, M., Loken, B., Fanzo, J., … Willett, W. (2018). Options for Keeping the Food System Within Environmental Limits. *Nature*, 562, 519–525. https://doi.org/10.1038/s41586-018-0594-0

Springmann, M., Godfray, H.C.J., Rayner, M., & Scarborough, P. (2016). Analysis and Valuation of the Health and Climate Change Cobenefits of Dietary Change. *PNAS*, 113(15), 4146–4151.

Tilman, D., & Clark, M. (2014). Global Diets Link Environmental Sustainability and Human Health. *Nature*, 515(7528), 518–522.

Tilman, D., Clark, M., Williams, D.R., Kimmel, K., Polasky, S., & Packer, C. (2017). Future Threats to Biodiversity and Pathways to Their Prevention. *Nature*, 546(7656), 73–81.

Toro-Mujica, P., García, A., Gómez-Castro, A., Perea, J., Rodríguez-Estévez, V., Angón, E., & Barba, C. (2012). Organic Dairy Sheep farms in South-Central Spain: typologies according to livestock management and economic variables. *Small Ruminant Research*, 104(1–3), 28–36. https://www.sciencedirect.com/science/article/pii/S0921448811004445

Tollefson, J. (2018). Clock Ticking on Climate Action. *Nature* (London), 562(7726), 172–173. https://doi.org/10.1038/d41586-018-06876-2

Toitū. (2020). *Summary of Toitu CarbonZero Certification: Fonterra Co-Operative Group Limited.* https://www.toitu.co.nz/__data/assets/pdf_file/0017/220922/Disclosure_1819_Fonterra_Enriched_Milk_CZ_Prod.pdf

Townsend-Small, A., & Hoschouer, J. (2021). Direct Measurements from Shut-in and Other Abandoned Wells in the Permian Basin of Texas Indicate Some Wells Are a Major Source of Methane Emissions and Produced Water. *Environmental Research Letters*, 16, 054081. doi:10.1088/1748-9326/abf06f

Tubb, C., & Seba, T. (2019). *Rethinking Food and Agriculture 2020–2030.* RethinkX Sector Disruption Report. https://www.rethinkx.com/food-and-agriculture

University of Otago Department of Human Nutrition. (2018). *Information Package for Users of the New Zealand Estimated Food Costs 2018* (Food Cost Survey 2018). http://hdl.handle.net/10523/8056

University of Otago and Ministry of Health. (2011). *A Focus on Nutrition: key findings of the 2008/09 New Zealand Adult Nutrition Survey*. New Zealand Ministry of Health. https://www.health.govt.nz/system/files/documents/publications/a-focus-on-nutrition-v2.pdf

Van Elswyk, M.E., & McNeill, S.H. (2014). Impact of Grass/Forage Feeding Versus Grain Finishing on Beef Nutrients and Sensory Quality: the U.S. experience. *Meat Science, 96*(1), 535–540. https://doi.org/10.1016/j.meatsci.2013.08.010

Weastell, L. (2020). *Conflict Between Intergenerational Family Farmers and Environmental Planning Processes: an 'economic versus environment' proposition or different ways of knowing?* [Unpublished doctoral thesis]. University of Canterbury. https://ir.canterbury.ac.nz/bitstream/handle/10092/101488/Weastell,%20Lynda_Final%20PhD%20Thesis.pdf?sequence=1

Whitmee, S., Haines, A., Beyrer, C., Boltz, F., Capon, A.G., de Souza Dias, B.F., Ezeh, A., Frumkin, H., Gong, P., Head, P., & Horton, R. (2015). Safeguarding Human Health in the Anthropocene Epoch: report of the Rockefeller foundation–*Lancet* Commission on planetary health. *The Lancet, 386*(10007), 1973–2028. https://doi.org/10.1016/S0140-6736(15)60901-1

Wiedemann, S., Davis, R., McGahan, E., Murphy, C., & Redding, M. (2016). Resource Use and Greenhouse Gas Emissions from Grain-Finishing Beef Cattle in Seven Australian Feedlots: a life cycle assessment. *Animal Production Science, 57*, 1149–1162. https://www.publish.csiro.au/an/an15454

Willett, W., Rockström, J., Loken, B., Springmann, M., Lang, T., Vermeulen, S., Garnett, T., Tilman, D., Declerck, F., Wood, A., Jonell, M., Clark, M., Gordon, L., Fanzo, J., Hawkes, C., Zurayk, R., Rivera, J., de Vries, W., Sibanda, L., & Murray, C. (2019). Food in the Anthropocene: the EAT–*Lancet* Commission on healthy diets from sustainable food systems. *The Lancet, 393*(10170), 447–492. https://doi.org/10.1016/S0140-6736(18)31788-4

Wittman, H. (2011). Food Sovereignty: a new rights framework for food and nature? *Environment and Society: Advances in Research, 2*, 87–105. doi:10.3167/ares.2011.020106

Governmental reports and technical notes

Australian Government. (2020). *Royal Commission into National Natural Disaster Arrangements Report.* https://naturaldisaster.royalcommission.gov.au/system/files/2020-11/Royal%20Commission%20into%20National%20Natural%20Disaster%20Arrangements%20-%20Report%20%20%5Baccessible%5D.pdf

Australian Institute of Health and Welfare. (2017). *Australia's Welfare 2017: in brief* (Cat. No. AUS 215). https://www.aihw.gov.au/reports/australias-welfare/australias-welfare-2017-in-brief/contents/our-working-lives

Australian Institute of Health and Welfare. (2019). *Poor Diet in Adults.* https://www.aihw.gov.au/reports/food-nutrition/poor-diet/contents/poor-diet-in-adults

Binks, B., Stenekes, N., Kruger, H., & Kancans, R. (2018). *Snapshot of Australia's Agricultural Workforce*. Australian Bureau of Agricultural and Resource Economics and Sciences. https://doi.org/10.25814/5c09cefb3fec5

Boulton, A., Kells, N., Beausoleil, N., Cogger, N., Johnson, C., Palmer., A., Laven, R., & O'Connor, A. (2018). *Bobby Calf Welfare Across the Supply Chain - Final Report for Ministry for Year 1* (MPI Discussion Technical Paper No. 2018/44). Ministry for Primary Industries, New Zealand Government. https://www.mpi.govt.nz/dmsdocument/30005-Bobby-Calf-Welfare-Across-the-Supply-Chain-Final-Report-Year-One-

British Nutrition Foundation. (1999). *Meat in the Diet*.

Climate Council of Australia. (2016, 16 October). *From Farm to Plate to the Atmosphere: food-related emissions*. https://www.climatecouncil.org.au/from-farm-to-plate-to-the-atmosphere-reducing-your-food-related-emissions/

Committee on Agriculture and Rural Development. (2019). *Report on the Proposal for a Regulation of the European Parliament and of the Council Amending Regulations (EU) No. 1308/2013*. European Parliament, Council of the European Union. https://www.europarl.europa.eu/doceo/document/A-8-2019-0198_EN.pdf#page=169

Department of Agriculture, Water and the Environment. (2020). *Australian Fisheries and Aquaculture Statistics*. Australian Government. https://www.agriculture.gov.au/abares/research-topics/fisheries/fisheries-data#australian-fisheries-and-aquaculture-statistics-201

Department of the Environment and Energy. (2019). *Carbon Neutral Program Public Disclosure Summary, The Northern Australian Pastoral Company*. Australian Government. https://napco.com.au/wp-content/uploads/2019/11/CNP_PDS_NAPCO2018v4.pdf

Environment Canterbury. (2002). *Nitrate Concentrations in Canterbury Groundwater – a review of existing data* (Report No. R02/17). https://docs.niwa.co.nz/library/public/ECtrR02-17.pdf

FAO. (2004). *The State of Food and Agriculture 2003–2004*. FAO, UN.

FAO. (2006a). *Livestock's Long Shadow: environmental issues and options*. FAO, UN. http://www.fao.org/3/a0701e/a0701e00.htm

FAO. (2014). *Research Approaches and Methods for Evaluating the Protein Quality of Human Foods* (Report of a FAO Expert Working Group). FAO, UN. https://www.fao.org/3/i4325e/i4325e.pdf

FAO, UN, IFAD, UNICEF, WFP, & WHO. (2017). *The State of Food Security and Nutrition in the World 2017: building resilience for peace and food security*. FAO, UN.

Global Nutrition Report. (2020). *Country Nutrition Profiles 2020*. https://globalnutritionreport.org/resources/nutrition-profiles/oceania/australia-and-new-zealand/

Hanson, C., & Abraham, P. (2010). *Nitrate Contamination and Groundwater Chemistry – Ashburton-Hinds Plain* (Report No. R10/143). Environment Canterbury.

Intergovernmental Panel on Climate Change (IPCC). (2019a). *IPCC Special Report on Climate Change and Land: an IPCC special report on climate change, desertification, land degradation, sustainable land management, food security, and greenhouse gas fluxes in terrestrial ecosystems.* [P.R. Shukla, J. Skea, E. Calvo Buendia, V. Masson-Delmotte, H.-O. Pörtner, D.C. Roberts, P. Zhai, R. Slade, S. Connors, R. van Diemen, M. Ferrat, E. Haughey, S. Luz, S. Neogi, M. Pathak, J. Petzold, J. Portugal Pereira, P. Vyas, E. Huntley, K. Kissick, M. Belkacemi, & J. Malley (Eds.)]. https://www.ipcc. ch/report/ar6/wg1/#SPM

IPCC. (2021). *Climate Change 2021: The Physical Science Basis. Contribution of Working Group I to the Sixth Assessment Report of the Intergovernmental Panel on Climate Change.* [V. Masson-Delmotte, P. Zhai, A. Pirani, S.L. Connors, C. Péan, S. Berger, N. Caud, Y. Chen, L. Goldfarb, M.I. Gomis, M. Huang, K. Leitzell, E. Lonnoy, J.B.R. Matthews, T.K. Maycock, T. Waterfield, O. Yelekçi, R. Yu, & B. Zhou (Eds.)]. Cambridge University Press. doi:10.1017/9781009157896

Keywood, M.D., Emmerson, K.M., & Hibberd, M.F. (2016). Atmosphere. In: *Australia State of the Environment 2016.* Department of the Environment and Energy, Australian Government. https://soe.environment.gov.au/theme/atmosphere, doi:10.4226/94/58b65c70bc372

Land Air Water Aotearoa (LAWA). (2021, 26 September). *LAWA River Water Quality National Picture Summary 2021.* https://www.lawa.org.nz/explore-data/river-quality/

Metcalfe, D., & Bui, E. (2016). Land: Soil: formation and erosion. In: *Australia State of the Environment 2016.* Department of the Environment and Energy, Australian Government. https://soe.environment.gov.au/theme/land/topic/2016/soil-formation-and-erosion, doi:10.4226/94/58b6585f94911

Ministry for the Environment & Stats NZ. (2018). *New Zealand's Environmental Reporting Series: our land 2018.* Available at www.mfe.govt.nz and www.stats.govt.nz

Ministry of Health. (2018). *2013/14 New Zealand Health Survey: nutrition.* New Zealand Government. https://www.health.govt.nz/our-work/populations/maori-health/tatau-kahukura-maori-health-statistics/nga-tauwehe-tupono-me-te-marumaru-risk-and-protective-factors/nutrition

Ministry for Primary Industries. (2012). *Pastoral Input Trends in New Zealand: a snapshot.* New Zealand Government. https://mpi.govt.nz/dmsdocument/4168-pastoral-input-trends-in-new-zealand-a-snapshot

Ministry for Primary Industries. (2019). *Primary Industries Workforce, 2019.* New Zealand Government. https://www.mpi.govt.nz/dmsdocument/29273-Primary-industries-workforce-fact-sheets

Ministry for Primary Industries. (2022). *Situation and Outlook for Primary Industries (SOPI) Data.* New Zealand Government. https://www.mpi.govt.nz/resources-and-forms/economic-intelligence/data/

New Zealand Government. (2019). *The New Zealand Government Aquaculture Strategy.* https://www.mpi.govt.nz/dmsdocument/15895-The-Governments-Aquaculture-Strategy-to-2025

Parliamentary Commissioner for the Environment. (2004). *Growing for Good: intensive farming, sustainability and New Zealand's environment.* https://www.pce.parliament. nz/media/1684/growing-for-good-full.pdf

Primary Sector Council. (2020). *Fit for a Better World: agriculture, food & fibre sector vision and strategic direction towards 2030*. Ministry for Primary Industries, New Zealand Government. https://www.mpi.govt.nz/dmsdocument/41046/direct

Sutton, K., Larsen, N., Moggre, G-J., Huffman, L., Clothier, B., Eason, J., & Bourne, R. (2018). *Opportunities in plant based foods – PROTEIN*. A Plant & Food Research report prepared for: Ministry Primary Industries and Plant & Food Research (SPTS No. 15748). Institute for Plant & Food Research, New Zealand.

Stats NZ. (2021a, 15 April). *Agricultural Production Survey: Fertilisers – nitrogen and phosphorus*. https://www.stats.govt.nz/indicators/fertilisers-nitrogen-and-phosphorus

Stats NZ. (2021b, 15 April). *Irrigated Land*. https://www.stats.govt.nz/indicators/irrigated-land

Articles

Ag Journal Staff Writers. (2020). Who Owns Australia. *Ag Journal*. https://www.weeklytimesnow.com.au/agribusiness/agjournal?nk=7103604c1c1c8cf919df979558635bde-1605825898

BBC News Staff Writers. (2021a, 26 May). Shell: Netherlands Court Orders Oil Giant to Cut Emissions. *BBC News*. https://www.bbc.com/news/world-europe-57257982

BBC News Staff Writers. (2021b, 10 December). Could Scotland See a 'Green Laird Gold Rush'? *BBC News*. https://www.bbc.com/news/uk-scotland-highlands-islands-59592218

Bottemiller Evich, H., & McCrimmon, R. (2021, 29 June). Biden Wants to Pay Farmers to Grow Carbon-Capturing Crops. It's Complicated. *Politico*. https://www.politico.com/news/2021/06/29/biden-climate-farmers-carbon-496843

Brainard, C. (2010, 29 March). Meat vs. Miles: coverage of livestock, transportation emissions hypes controversy. *Columbia Journalism Review*. https://archives.cjr.org/the_observatory/meat_vs_miles.php

Burwood-Taylor, L. (2019, 10 September). Kraft Heinz's VC Invests in New Culture $3.5m Seed Round for Cell-Grown Cheese. *Ag Funder News*. https://agfundernews.com/brief-kraft-heinz-invests-in-new-culture-3-5m-seed-round-for-cell-grown-cheese

Business Wire Staff Writers. (2020, 1 December). Eat Just Granted World's First Regulatory Approval for Cultivated Meat. *Business Wire*. https://www.businesswire.com/news/home/20201201006251/en/Eat-Just-Granted-World%E2%80%99s-First-Regulatory-Approval-for-Cultured-Meat

Byrne, J. (2020, 2 October). Academic Rails Against 'Flawed' Analogy Comparing EU Livestock Emissions to That of Cars and Vans. *Feed Navigator*. https://www.feednavigator.com/Article/2020/10/02/Academic-rails-against-flawed-analogy-comparing-EU-livestock-emissions-to-that-of-cars-and-vans

Campos, A., Wasley, A., Heal, A., Phillips, D., & Locatelli, P. (2020, 27 July). Revealed: New Evidence Links Brazil Meat Giant JBS to Amazon Deforestation. *The Guardian*. https://www.theguardian.com/environment/2020/jul/27/revealed-new-evidence-links-brazil-meat-giant-jbs-to-amazon-deforestation

Farm

Chan, D. (2019, 31 July). We're Finally Cracking Vegan Cheese's Terrible Taste Problem. *Wired*. https://www.wired.co.uk/article/vegan-cheese-uk-recipe-taste#:~:text=%E2%80%9CFor%20plant%2Dbased%20cheeses%2C,their%20comapany%20in%20January%202019

Chef's Pencil Staff Writers. (2019, 16 January). The Most Popular Countries and Cities for Vegans in 2018. *Chef's Pencil*. https://www.chefspencil.com/where-veganism-is-most-popular-around-the-world-in-2018/

Condon, J. (2021, 6 January). Australian Beef Exports Fall 15pc, to 1.039 Million Tonnes. *Beef Central*. https://www.beefcentral.com/trade/2020-beef-exports-fall-15pc-to-1-039-million-tonnes/

Draper, R. (2009, April). Australia's Dry Run. *National Geographic*. https://www.nationalgeographic.com/magazine/2009/04/murray-darling/

Elgin, B. (2020, 9 December). These Trees Are Not What They Seem: how the Nature Conservancy, the world's biggest environmental group, became a dealer of meaningless carbon offsets. *Bloomberg*. https://www.bloomberg.com/features/2020-nature-conservancy-carbon-offsets-trees

Farm Online Staff Writers. (2019). Nats Name Labelling Laws 'Number One' Issue in Food Regulation. *Farm Online National*. https://www.farmonline.com.au/story/6389156/nats-name-labelling-laws-number-one-issue-in-food-regulation/

Franck, T. (2019, 23 May). Alternative Meat to Become $140 Billion Industry in a Decade, Barclays Predicts. *CNBC*. https://www.cnbc.com/2019/05/23/alternative-meat-to-become-140-billion-industry-barclays-says.html

Friedman, L. (2021, 2 November). Biden Administration Moves to Limit Methane, A Potent Greenhouse Gas. *New York Times*. https://www.nytimes.com/2021/11/02/climate/biden-methane-climate.html

Ghosh, I. (2020, 15 May). Zoom is Now Worth More Than the World's 7 Biggest Airlines. *Visual Capitalist*. https://www.visualcapitalist.com/zoom-boom-biggest-airlines/

Gibson, E. (2019, 27 February). Fonterra Dips Hoof in Alternative Animal Products. Newsroom. https://www.newsroom.co.nz/fonterra-dips-hoof-in-alternative-animal-products

Grain Staff Writers. (2021, 21 September). Big Food in Africa: endangering people's health. *Grain*. https://grain.org/en/article/6724-big-food-in-africa-endangering-people-s-health

Gunther, M. (2016, 6 April). New Crop – the vegan venture fund fighting for animal rights. *The Guardian*. https://www.theguardian.com/sustainable-business/2016/apr/06/new-crop-capital-vegetarianism-vegan-animal-rights-agriculture-usda-beyond-meat

Hamblin, J. (2017, 2 August). If Everyone Ate Beans Instead of Beef: with one dietary change, the U.S. could almost meet greenhouse-gas emissions goals. *The Atlantic*. https://www.theatlantic.com/health/archive/2017/08/if-everyone-ate-beans-instead-of-beef/535536/

Jamieson, A., & Boyle, A. (2013, 6 August). 'Intense Flavor': the $330,000 burger that was built in a lab hits the spot. *NBC News*. https://www.nbcnews.com/technolog/intense-flavor-330-000-burger-was-built-lab-hits-spot-6c10835460

Johnson, A. (2020). The Role of Red Meat in Healthy & Sustainable New Zealand Diets. *Beef and Lamb New Zealand*. https://www.beeflambnz.co.nz/how-much-red-meat-are-we-eating

Keane, P. (2020, 20 September). How the Oil Industry Made Us Doubt Climate Change. *BBC News*. https://www.bbc.com/news/stories-53640382

Kembrey, M. (2019, 8 March). Lunch with: Artist Janet Laurence. *Sydney Morning Herald*. https://www.smh.com.au/entertainment/art-and-design/lunch-with-artist-janet-laurence-20190304-h1byv0.html

Kitchin, T. (2020, 24 June). Hawke's Bay Could Become Desert if Climate Change Ignored - Regional Council. *Radio New Zealand*. https://www.rnz.co.nz/news/national/419763/hawke-s-bay-could-become-desert-if-climate-change-ignored-regional-council

Lawrence, F. (2005, 8 December). Multinationals, Not Farmers, Reap Biggest Rewards in Britain's Share of CAP Payouts. *The Guardian*. https://www.theguardian.com/uk/2005/dec/08/freedomofinformation.foodanddrink

Lawton, N. (2018, 8 August). Protestors Gather at Tegel HQ in Auckland to Oppose Mega Chicken Farm Plans. *Stuff*. https://www.stuff.co.nz/auckland/106034297/protesters-gather-at-tegel-hq-in-auckland-to-oppose-mega-chicken-farm-plans

Levitt, T. (2018, 26 March). Dairy's 'Dirty Secret': it's still cheaper to kill male calves than to rear them. *The Guardian*. https://www.theguardian.com/environment/2018/mar/26/dairy-dirty-secret-its-still-cheaper-to-kill-male-calves-than-to-rear-them

Levitt, T. (2020, 10 December). The End of Dairy's 'Dirty Secret'? farms have a year to stop killing male calves. *The Guardian*. https://www.theguardian.com/environment/2020/dec/10/the-end-of-dairys-dirty-secret-farms-have-a-year-to-stop-killing-male-calves

Marshall, A.R.C. (2022, 27 January). Who Owns Scotland: the rise of the green lairds. *Reuters*. https://www.reuters.com/investigates/special-report/scotland-environment-green-lairds/

Mitloehner, F.M. (2018, 25 October). Yes, Eating Meat Affects the Environment, but Cows Are Not Killing the Climate. *The Conversation*. https://theconversation.com/yes-eating-meat-affects-the-environment-but-cows-are-not-killing-the-climate-94968

Morton, A. (2021, 27 May). Australian Court Finds Government Has Duty to Protect Young People from Climate Crisis. *The Guardian*. https://www.theguardian.com/australia-news/2021/may/27/australian-court-finds-government-has-duty-to-protect-young-people-from-climate-crisis

Mottet, A., & Steinfeld, H. (2018, 16 September). Cars or Livestock: which contribute more to climate change? *Thomson Reuters Foundation News*. https://news.trust.org/item/20180918083629-d2wf0

New Zealand Herald Staff Writers. (2007, 8 February). Drought Hit Australian Farmers to Holiday in NZ. *The New Zealand Herald*. https://www.nzherald.co.nz/nz/drought-hit-australian-farmers-to-holiday-in-nz/JPC3TLVS637A3QUWNZ73RE45NA/

Farm

Nowland, D. (2017, 15 December). The Truth About Veal. *Jamie Oliver Online*. https://www.jamieoliver.com/features/the-truth-about-veal/

Phillips, M. (2021, 9 June). Exxon's Board Defeat Signals the Rise of Social-Good Activists. *New York Times*. https://www.nytimes.com/2021/06/09/business/exxon-mobil-engine-no1-activist.html

Philpott, T. (2021, 2 August). Is Lab Meat About to Hit Your Dinner Plate? *Mother Jones*. https://www.motherjones.com/food/2021/08/is-lab-meat-about-to-hit-your-dinner-plate

Plant Based News Staff Writers. (2019, 15 March). Vegan Kids Join Crowd of 300,000 in Australian Climate Strike. *Plant Based News*. https://plantbasednews.org/news/vegan-kids-join-300-000-australian-climate-strike/

Purdy, C. (2016, 13 November). How the Vegan Movement Broke Out of Its Echo Chamber and Finally Started Disrupting Things. *Quartz*. https://qz.com/829956/how-the-vegan-movement-broke-out-of-its-echo-chamber-and-finally-started-disrupting-things/

Sheikh, K. (2019, 2 August). Got Impossible Milk? the quest for lab-made dairy. *New York Times*. https://www.nytimes.com/2019/08/02/science/lab-grown-milk.html

Strachan, G. (2020, 7 July). India's Golden Bridge is Still Standing 139 Years After Being Made from the Same Iron As the Doomed First Tay Rail Bridge. *The Courier Evening Telegraph*. https://www.thecourier.co.uk/fp/business-environment/transport/1419719/indias-golden-bridge-is-still-standing-139-years-after-being-made-from-the-same-iron-as-the-doomed-first-tay-rail-bridge/

Ritchie, H. (2020, 10 March). The Carbon Footprint of Foods: are differences explained by the impacts of methane? *Our World in Data*. https://ourworldindata.org/carbon-footprint-food-methane

Verleg, A., Bhole, A., Logan, T., Burton, L. (2019, 8 June). Carbon Neutral Livestock Production — consumers want it and farmers say it is achievable. *ABC News*. https://www.abc.net.au/news/rural/2019-06-08/carbon-neutral-livestock-achievable-by-2030-says-mla/11046592

Wasley, A., Harvey, F., Davies, M., & Child, D. (2017, 17 July). UK Has Nearly 800 Livestock Mega Farms, Investigation Reveals. *The Guardian*. https://www.theguardian.com/environment/2017/jul/17/uk-has-nearly-800-livestock-mega-farms-investigation-reveals

Xia, L. (2021, 14 September). What Toilet-Trained Cows Could Mean for Reducing Greenhouse Gas Emissions. *Stuff*. https://www.stuff.co.nz/national/126361447/what-toilettrained-cows-could-mean-for-reducing-greenhouse-gas-emissions

Zimberoff, L. (2019, 11 July). Forget Synthetic Meat, Lab Grown Dairy is Here. *Bloomberg*. https://www.bloomberg.com/news/articles/2019-07-11/forget-cultured-meats-lab-grown-dairy-is-attracting-investors#xj4y7vzkg

Press releases and corporate statements

Beef and Lamb NZ. (2021). *Farm Facts 45th Edition* (Pub No. P21002). https://beeflambnz.com/sites/default/files/data/files/Compendium%202021_digital.pdf

Carlson, R. (2022, 10 February). Nori's Stance on Soil Sampling: why soil sampling isn't (yet) a silver bullet for soil carbon credits. *Medium.* https://medium.com/nori-carbon-removal/noris-stance-on-soil-sampling-540e30f1655d

Dairy Australia. (2020). *Bobby Calves* [Technical note]. https://www.dairyaustralia.com.au/animal-management-and-milk-quality/calf-rearing/bobby-calves#.Yf8tZFhByqR

FAO. (2006b, 29 November). *Livestock a Major Threat to Environment. Remedies Urgently Needed* [Press release].

Future Beef Knowledge Centre. (2011, 16 September). *Feedlots.* https://futurebeef.com.au/knowledge-centre/feedlots/

Gordon, W., Gantori, S., Gordon, J., Leemann, R., & Boer, R. (2019). The Food Revolution: the future of food and the challenges we face. *UBS.* https://www.ubs.com/global/en/wealth-management/chief-investment-office/investment-opportunities/sustainable-investing/2019/food-revolution.htm

Impossible Foods. (2018). *Mission: Earth Campaign* (3rd Annual Shorty Social Good Awards application). https://shortyawards.com/3rd-socialgood/impossible-foods-mission-earth-campaign

Impossible Foods. (2019). *Regenerative Grazing: the 'clean coal' of beef* (Impact Report 2019). https://impossiblefoods.com/mission/2019impact/grassfedbeef/

Innova Market Insights. (2021). What, Where and How? consumers want food transparency. https://www.innovamarketinsights.com/blog/what-where-and-how-consumers-want-food-transparency/

IPCC. (2019b, 8 August). *Land is a Critical Resource, IPCC Report Says* [Press release]. https://www.ipcc.ch/2019/08/08/land-is-a-critical-resource_srccl/

Label Insight. (2018). *The Transparency Imperative: product labelling from the consumer perspective.* https://www.fmi.org/forms/store/ProductFormPublic/the-transparency-imperative-product-labeling-from-the-consumer-perspective

Meat Industry Association of New Zealand (MIA). (2021a, 4 February). *New Zealand Red Meat Exports Reached Historic High Levels During 2020* [Press release]. https://www.mia.co.nz/news-and-views/new-zealand-red-meat-exports-reached-historic-high-levels-during-2020/

MIA. (2021b). *Annual Report 2021.* https://www.mia.co.nz/assets/Uploads/MIA-Annual-Report-2021-Final.pdf

Meat and Livestock Australia (MLA). (2020). *State of the Industry Report.* https://www.mla.com.au/globalassets/mla-corporate/prices--markets/documents/trends--analysis/soti-report/mla-state-of-industry-report-2020.pdf

Mintel. (2020, 17 January). *Plant-Based Push: UK sales of meat-free foods shoot up 40% between 2014–19* [Press release]. https://www.mintel.com/press-centre/food-and-drink/plant-based-push-uk-sales-of-meat-free-foods-shoot-up-40-between-2014-19

Farm

Otis Oat Milk. (2021). *Otis Carbon Reduction Plan 2021*. https://cdn.shopify.com/s/files/1/0419/8639/2219/files/Otis_Oat_Milk_CRP_5ac5aee5-279b-4ae7-b57b-56831b0e1d24.pdf?v=1636505284

Price Waterhouse Cooper (PWC). (2011). *The Australian Beef Industry: from family farm to international markets*. https://www.pwc.com.au/industry/agribusiness/assets/australian-beef-industry-nov11.pdf

Taylor, A. (2021, 21 October). *Cutting Emissions the Australian Way* (Hon Angus Taylor MP Opinion Piece). https://www.minister.industry.gov.au/ministers/taylor/opinion-piece/cutting-emissions-australian-way

Thorbecke, M., & Dettling, J. (2019). *Carbon Footprint Evaluation of Regenerative Grazing at White Oak Pastures*. Qantis. https://blog.whiteoakpastures.com/hubfs/WOP-LCA-Quantis-2019.pdf

UN. (2006, 29 November). Rearing Cattle Produces More Greenhouse Gases Than Driving Cars, UN Report Warns. *UN News*. https://news.un.org/en/story/2006/11/201222-rearing-cattle-produces-more-greenhouse-gases-driving-cars-un-report-warns

University of Cincinnati. (2021, 28 April). Inactive Oil Wells Could Be Big Source of Methane Emissions: geologist studies greenhouse gas emissions from uncapped, idle wells in Texas. *ScienceDaily*. www.sciencedaily.com/releases/2021/04/210428132941.htm

The White House. (2021, 18 September). *Joint US-EU Press Release on the Global Methane Pledge* [Press release]. https://www.whitehouse.gov/briefing-room/statements-releases/2021/09/18/joint-us-eu-press-release-on-the-global-methane-pledge/

Woolworths Group. (2021). *Animal Welfare Policy*. https://www.woolworthsgroup.com.au/icms_docs/195961_woolworths-group-animal-welfare-policy.pdf

WSP. (2021). *ISO-Conformant Report: comparative Life Cycle Assessment of Perfect Day whey protein production to dairy protein*. Perfect Day Inc. https://m4f6w9b2.rocketcdn.me/app/uploads/2022/01/Comparative-Perfect-Day-Whey-LCA-report-prepared-by-WSP_20AUG2021_Non-Confidential-1.pdf

WWF. (2020). *Living Planet Report 2020: bending the curve of biodiversity loss*. [R.E.A. Almond, M. Grooten, & T. Petersen (Eds.)]. https://www.zsl.org/sites/default/files/LPR%202020%20Full%20report.pdf

Audiovisual media

Bascomb, B., & Doering, J. (2020, 25 September). *A Win for Native American Sovereignty*. Living on Earth [Radio broadcast transcript]. https://www.loe.org/shows/segments.html?programID=20-P13-00039&segmentID=3

Harvey, N., & Brown, N. (Writers and Producers). (2019). *A Carnivore's Crisis* [Podcast]. Audible Originals, Pipi Films & Duststar.

Rapley, R. (Director). (2020). *The Man Who Tried to Feed the World* [Documentary]. PBS. https://www.pbs.org/wgbh/americanexperience/films/man-who-tried-to-feed-the-world/#film_description

Tippet, K. (Host). (2018, 3 May). Nature, Joy, and Human Becoming [Podcast episode]. In *On Being with Krista Tippet* [Podcast]. WNYC Studios. https://onbeing.org/programs/michael-mccarthy-nature-joy-and-human-becoming/

Tippet, K. (Host). (2022). *On Being with Krista Tippet* [Podcast]. WNYC Studios.

Westervelt. (Host). (2020). Season Three, *Drilled: a true crime podcast about climate change* [Podcast]. Critical Frequency.

Endnotes

Introduction

1   For a broad review see Goodman & Redclift (1991), p. 120. And for specific information, see Pingali (2012).

2   The seed modification program was led by American agronomist and Nobel Laureate Norman Borlaug. For more information read Hesser (2019), or watch Rapley (2020).

3   Between 1960 and 2000, wheat yields in all developing countries rose 208 per cent. Maize corn jumped 157 per cent and rice more than 100 per cent. These figures are seen as proof of success of the Green Revolution technology. Elsewhere researchers have highlighted the failures of the scheme especially on the African and Indian continents, pinpointing soil degradation from monocropping, heavy water use, and high input costs for indebted farmers as key concerns. See FAO (2004).

4   Researchers have shown that the dominance of these four crops at the expense of endemic crops and vegetables diminished community food resilience and led to a reliance on imported manufactured food purchased in the supermarket. Some consider the Green Revolution an extension of the colonial project. See Pingali (2012); Altieri & Pengue (2006).

5   Gómez et al. (2013).

6   Khoury et al. (2014).

7   Fontefrancesco (2019).

8   FAO (2006a).

9   FAO (2006b).

10 In 2010, Dr Frank Mitloehner, an air quality scientist from the University of California, Davis, delivered a talk based on a peer-reviewed paper asserting that the FOA's comparison modelling between livestock and transport was flawed. He said that while livestock's emissions were based on the whole lifecycle — every step along the production chain was accounted for, from cattle burps to the making and application of fertiliser — transport's emissions were not. In the wake of Dr Mitloehner's criticisms of the FOA report he told *Columbia Journalism Review* that the media's single-minded focus on conflict between him and the FOA was unfortunate because despite the faulty comparison, the FAO was doing valuable research. See Brainard (2010).

1. Big bad beef

1   MIA (2021a).
2   Condon (2021).
3   PWC (2011).
4   MIA (2021b).
5   Gow (2014).
6   Environment Canterbury (2002); Hanson & Abraham (2010).
7   The impacts of the subsidy loss are covered extensively in Wallace (2014).
8   Stats NZ (2021a).
9   Stats NZ (2021b).
10 Maseyk (2021).
11 IPCC (2019a).
12 IPCC (2019b).
13 MLA (2020).
14 Johnson (2020).
15 OECD (2020), Meat Consumption (indicator) available at https://doi.org/10.1787/fa290fd0-en
16 Harvey & Brown (2019). Original interview with Sophie Egan conducted October 2019 by Naima Brown. Point further clarified via email correspondence with Nicola Harvey in November 2020.
17 Springmann et al. (2018).
18 Willett et al. (2019).
19 Springmann et al. (2018)
20 The White House (2021).
21 University of Cincinnati (2021).
22 Friedman (2021).
23 Keywood et al. (2016).
24 Allen (2015).
25 In email correspondence from October 2021, Minister for the Environment James Shaw wrote: 'It is worth noting that the current use of the GWP-100 metric was settled only in 2019, after arduous international negotiations. There would be almost no appetite internationally to re-open this negotiation, and even if we did so, it would provide impetus for others to try to re-open other parts of the negotiations.
26 Interview with Professor Frame was conducted via Zoom on 2 November 2020. Further information quoted in this chapter was received via email on 11 December 2020.
27 Elgin (2020).
28 The successful Scottish craft beer brewery BrewDog is one of many companies now referred to as 'green lairds', due to its purchase of 9,000 acres of land in the Scottish Highlands to be converted into permanent forest for carbon offsets. For more information see Marshall (2022) and BBC News Staff Writers (2021b).
29 Taylor (2021).
30 The progeny from the North Australian Pastoral Company's breeding herd typically end up in a 18,000-head feedlot on the border of Queensland and New South Wales. The company was given carbon neutral accreditation by the Australian Federal

Government in 2019. The company's Commercial General Manager Stephen Moore has said that the off-setting is just a starting point. And in time the company will focus on 'the animal' and do things like providing feed that will decrease methane emissions. The company has plans to convert diesel-powered bores to solar, and engage in feed planting regimes that will help the cattle get fatter faster, thereby lessening their time on this planet. For more information see Verleg et al. (2019).

31   Department of the Environment and Energy (2019).

2. Silver linings, silver bullets

1   Ministry for Primary Industries (2022).
2   Dairy Australia (2020).
3   Boulton et al. (2018).
4   Levitt (2018); Levitt (2020).
5   Nowland (2017).
6   Harker-Schuchet al. (2018).
7   Plant Based News Staff Writers (2019).
8   IPCC (2021).
9   Harwatt (2019).
10   Willett, et al. (2019).
11   This theme is addressed extensively in Friedmann (2005).
12   Friedmann's checklist is referenced in Campbell (2019).
13   This theme is referenced repeatedly in Grace (2018).
14   See Jane Dixon's *The Changing Chicken: Chooks, Cooks and Culinary Culture* (2002) referenced in Parker et al. (2019).
15   Parker et al. (2019).
16   See Australian Eggs, *What Are Free-Range Eggs*, available at https://www. australianeggs.org.au/farming/free-range-eggs
17   Woolworths Group (Updated 2021),
18   See Jane Dixon's *The Changing Chicken: Chooks, Cooks and Culinary Culture* (2002) referenced in Parker et al. (2019).
19   Parker et al. (2019).
20   Carey et al. (2017).
21   Chef's Pencils Staff writers (2019).
22   Mintel (2020).
23   Kidd et al. (2021).
24   Harvey & Brown (2019). Interview with Sophie Egan conducted October 2019 by Naima Brown.
25   Franck (2019); Gordon et al. (2019).
26   Foley et al. (2011); Carlson et al. (2016).
27   Potter et al. (2020).
28   Ritchie (2020).
29   Potter et al. (2020).
30   Hegab et al. (2014).
31   Ministry for Primary Industries (2019); Binks et al. (2018).
32   FAO (2014).

33 Van Elswyk & McNeill (2014).

34 Little (2019).

35 Burwood-Taylor (2019).

36 Email correspondence with Fonterra Communications (15 May 2021).

37 Gibson (2019).

38 Metcalfe & Bui (2016).

39 Ministry for the Environment & Stats NZ (2018).

40 LAWA (2021).

41 Metcalfe & Bui (2016).

42 Otis Oat Milk (2021).

43 Toitū (2020).

44 A comment that references the formative 1987 decision by Labour prime minister David Lange's government to ban US warships carrying nuclear weapons from entering New Zealand waters. It was seen as a diplomatic and legislative protest in support of the rising public opinion against nuclear testing in the Pacific. The country has been nuclear free since.

45 Interview with Professor Carolan conducted via Zoom, November 2019.

46 Interview with Dr Louise Heath conducted via phone, 4 October 2019.

47 Studies indicate that more than 60 per cent of the energy source in subsistence diets among twentieth century hunter-gatherers came from animal foods. What Dr Heath is arguing is that small animals and insects were an essential component of early hunter-gatherer diets, but that these were sourced by woman. Interview conducted via phone, 4 October 2019. Further findings available in Cordain et al. (2002).

48 Beck et al. (2012); Cook et al. (1994).

49 Caffeine contains polyphenols that can latch on to iron, working to reduce absorption when in the digestive system. For more information see Ma et al. (2011).

50 Graham (2017).

51 In a small survey the Auckland City Mission (2019) estimated that ten per cent of New Zealand's population have experienced food insecurity. An earlier report by University of Otago & Ministry of Health (2011) put the figure closer to 40 per cent of adults and 20 per cent of children. And the percentage is growing each year. In Australia it is estimated that up to 13 per cent of the population are food insecure, a number that more than doubles for Indigenous and Torres Strait Islander peoples. See Bowden (2020).

52 Carter et al. (2010).

3. Trust in waste

1 Australian Institute of Health and Welfare (2017).

2 Climate Council of Australia (2016).

3 Newman et al. (2020).

4 Lawrence (2005).

5 Committee on Agriculture and Rural Development (2019).

6 Farm Online Staff Writers (2019).

7 Ibid

8 Jamieson & Boyle (2013).

9   Business Wire Staff Writers (2020).

10   Philpott (2021.

11   Gunther (2016).

12   Innova Market Insights (2021).

13   Label Insight (2018).

14   Campos et al. (2020).

15   Sutton et al. (2020).

16   Ibid

17   Carolan (2018)..

18   Email correspondence with Countdown Communications team, 15 July 2021.

19   Interview with Professor Carolan conducted October 2019.

20   Schulman (2018.

21   Carolan (2018).

22   Ibid, p185

23   Pereira et al. (2020).

4. Corn fritters in the colonies

1   Ibid, p. 8.

2   Ibid.

3   National Trust (2020).

4   Strachan (2020).

5   Nussbaum (2011) p. 105.

6   Archival letter stamped Morvi, India. Dated 26 October 1901. Addressed to CP White Esq, manager and engineer in chief. Signed by 12 staff of the Morvi railway.

7   Draper (2009).

8   Anthony (2007).

9   Gammage, (2011).

10   New Zealand Herald Staff Writers (2007).

11   Friedmann (2005).

12   Primary Sector Council (2020).

13   Campbell (2009).

14   Macgregor (1981), p. 215.

15   Description sourced from self-published memoir of Everard White.

5. Meat money

1   Australian Government (2020).

2   Evans (2019).

3   Craig (1974).

4   Moriarty et al. (2011).

5   Ghosh (2020).

6   Little (2019), p. 148.

7   Tippet (2018). Quote sourced from an interview with Michael McCarthy.

8   Future Beef Knowledge Centre (2011).

9   Harvey & Brown (2019). Interview conducted by Naima Brown.

Farm

10 Interview with Robert Pekin conducted via Zoom, August 2020.

11 Garnett et al. (2017).

12 Wiedemann et al. (2016).

13 Wasley et al. (2017).

14 Lawton (2018).

15 Bascomb & Doering (2020), interview with Rebecca Nagel.

16 Ag Journal Staff Writers (2020).

17 Kembrey (2019).

6. Shortcuts

1   Evans (2021).

2   Harvey & Brown (2019). Interview conducted October 2019 by Hayley Redman.

3   Kennedy et al. (2020).

4   Interview with Stuart Austin conducted by phone, April 2021.

5   See Twitter thread from February 12 2022. https://twitter.com/paigestanley_ag/status/1492343501216190469?s=20&t=nzqyJImyebt3IK0AuQSgiQ

See also Carlson (2022).

6   For more information about the subsidy and payment schemes visit: https://www.cdfa.ca.gov/healthysoils/

7   Bottemiller Evich & McCrimmon (2021).

8   Harvey & Brown (2019). Interview with Joe Towers conducted October 2019 by Hayley Redman. Further email correspondence with Joe and Edward Towers throughout 2021.

9   Grelet et al. (2021).

10 Kinley et al. (2016).

11 Lind (2013), p. 169.

12 Ibid p. 147.

13 Interview with Professor Frank Griffin conducted August 2021.

14 Hafting et al. (2012).

15 This is the type of innovation lead authors at the IPCC are calling for with the aim to intensify and upscale low-energy protein farming: fin fish farms surrounded by seaweed mean greater volumes of protein not sourced from methane-emitting cattle. The wild fish industry, long governed by a quota system, pulls approximately 600,000 tonnes of fish from the waters around New Zealand each year, with almost half going into the export market to generate $NZ1.8 billion in earnings. In Australia, the fisheries, both wild and aquaculture, earn more than 3 billion dollars. Dominated by Tasmania and Queensland, the sector has been steadily growing in value over the past decade, even as wild fishery production remains constant. The export demand for salmon, tuna, and lobster continues to push the sector forward towards aquaculture. More than half of the seafood consumed globally is farmed, and the New Zealand government wants a 3-billion-dollar industry by 2035, including land and open ocean farms. In a strategy document released in 2019, the New Zealand Labour government highlights the efficiency of aquaculture, forecasting that the annual value for every ten farmed hectares of salmon is $NZ140,000,000, versus $77,000 for dairy and $8,500 for sheep and beef. The profit margins and efficiency forecasts matter. Farmable land

is limited, so the future of food production is focused on intensifying but without the harmful environmental footprint: i.e., sustainable fisheries. It's a delicate balance.

See Department of Agriculture, Water and the Environment (2020); New Zealand Government (2019).

16 Impossible Foods (2019).

17 Harvey & Brown (2019). Interview with Nicolette Hahn Niman conducted October 2019.

18 Ibid.

19 Impossible Foods (2018).

20 Harvey & Brown (2019). Interview with Josh Tetrick conducted October 2019 by Naima Brown.

21 Ibid.

22 Lynch & Pierrehumbert (2019).

23 Diary processing has become so efficient that there's barely any waste product. You may think of milk, cheese, and yoghurt as the mainstays, but milk also becomes usable sugars, proteins, and fats in the way of lactose, whey, and anhydrous milk fat. These can be extracted, packaged, and sold to manufacturers for use in products like pharmaceuticals, crisp flavouring, bread, pasta, confectionary, chocolate, and the list goes on. .

24 WSP (2021).

25 Mazzetto, et al. (2021).

26 Allen et al. (2018).

27 Ibid.

28 Westervelt (2020); Keane (2020).

7. Change maker

1 Xia (2021).

2 Nawroth et al. (2019).

3 Austin, et al. (2005).

4 Montgomery (2013)/

5 Porcher, (2006).

6 WWF (2020/

7 Wilkerson (2020), p. 17.

8 Perssonet al. (2020).

9 Bartram & Baldwin (2010).

10 Leibler et al. (2017).

11 Richards et al. (2013); Grandin (1998); Dillard (2008).

12 The 2020 online petition 'Help stop meatworks from working and keeping them safe at home with there family's through this time' [sic] is available at https://www.change.org/p/government-meatworks-to-stopping-working

8. Global and local

1 Weastell (2020).

2 Pereira et al. (2019).

Farm

3   Interview with Amy Mcquire conducted October 2020.
4   Fiso (2020), p. 65.
5   Interview with Dr Lisa Te Morenga conducted via phone, July 2021.
6   Ministry of Health (2018).
7   Australian Institute of Health and Welfare (2019).
8   Global Nutrition Report. (2020).
9   Rush & Obolonkin (2020).
10 Kinnunen et al. (2020).
11 Khoury et al. (2014).
12 Pereira et al. (2020).
13 Bell (2011).
14 Wittman (2011).
15 Interview with Dr Jessica Hutchings conducted via phone, November 2020.
16 Lang et al. (2017).
17 See Red Tractor infographics at: https://twitter.com/redtractorfood/status/1405181439872704516?lang=en
18 See tweet from Rhian Price, *Farmer's Weekly* Livestock Editor at https://twitter.com/RhianepFW/status/1247611032526368770

9. Don't give up

1   Kitchin (2020).
2   Morton (2021).
3   BBC News Staff Writers (2021a).
4   Phillips (2021).
5   This sentiment was extolled by Burkeman (2021) in *Four Thousand Weeks*, and during an interview with Tippet (2022).